DEATH EXPLAINED

A Ghost Hunter's Guide to the Afterlife

By
Michael Dupler

with Michelle Dupler

A Shot in the Dark Media

A Shot in the Dark Media
Columbus, Ohio 43204

www.ashotinthedarkmedia.com

Copyright © 2013 by Michael Dupler

Cover design by Michelle Dupler

Book design by Michelle Dupler

All rights reserved. This book or any portion thereof may not be reproduced or used in any manner whatsoever without the express written permission of the publisher except for the use of brief quotations in a book review.

Printed in the United States of America

First Printing, July 2013

ISBN 978-0-9897112-0-3

TABLE OF CONTENTS

Introduction: What's it all about? 7

Chapter 1: Adventures of a Fledgling Ghost Hunter 10

Chapter 2: Haunting Theories 24

Chapter 3: A Journey Into Humanity's Past 31

Chapter 4: God and Ghosts 50

Chapter 5: Confronting Death 65

Chapter 6: What's Physics Got To Do With It? 71

Chapter 7: A Symphony of Strings 84

Chapter 8: Into the Wormhole 93

Chapter 9: Time in a Bottle 99

Chapter 10: Pulling It All Together 104

Chapter 11: Why Hauntings? 111

Chapter 12: The Case for Scientific Inquiry 121

Conclusion 126

Epilogue: Ghost Hunting 101 129

    Ghost Hunting Equipment 129

    Places to Shop Online 136

    Haunted Locations 138

Bibliography 154

*Dedicated to my grandchildren — Arianna and Brennan.
I encourage you to think for yourselves.*

INTRODUCTION
What's it all about?

My journey into the world of the paranormal began several years ago when I started watching the *Ghost Hunters* series on the SyFy Channel featuring paranormal investigations by The Atlantic Paranormal Society, or TAPS. I initially started watching just for entertainment. I've always had a fascination with the unexplained, whether it's the possibility of ghosts, or aliens among us, or cryptids like Bigfoot. I'm drawn to the mysteries of the universe and the possibility there may be more to life and the world than what we see every day. But as I entered my 50s and became more conscious of my own mortality and the inevitability of my death, I became particularly intrigued by what the Ghost Hunters crew were doing — and with questions about what happens to us after we die. I kept in mind that this was just a TV show, but I also began to wonder if ghosts really did exist. Was there really life after death?

An epiphany came one night a few years ago while I was watching the show. The TAPS team was investigating an abandoned train station in Buffalo, New York, where the stories about ghostly activity included claims that apparitions of people would gather around a water fountain on one of the floors — but no water fountain is there. On another floor, the previous owner had an apartment where visitors at times would claim to hear voices and have creepy feelings. In the rotunda area, shadow figures could be seen moving from one door to another. In a tunnel entrance to the basement area, a worker was said to hear a voice speaking German. During World War II, German prisoners of war were used as labor for maintenance and cleaning projects.

During the investigation, they used a thermal imaging camera to try to pick up any anomalous heat signatures. At the end of the hallway they caught a heat source that they couldn't explain. While they were there, they heard a woman laughing and footsteps that seemed to walk with them in another room. They went to the previous owner's apartment where they encountered high EMF[1] readings. While they were there, they heard footsteps and a loud bang. At one time, they heard a loud voice say, "Go home!" This was loud enough to hear and to catch on their voice recorders. As they reviewed the evidence gathered during the investigation, they also caught an EVP[2] from that area that sounded like it said, "Bring the girls." As it so happens, there were two female investigators with the team that night.

While the investigators were in the main concourse area on the first floor, they started getting readings from their K2 unit. A K2 is an electromagnetic field detector like an EMF detector,

---

[1] EMF stands for electromagnetic fields and can be registered on a Gauss meter or what is called an EMF detector.

[2] EVP stands for electronic voice phenomena — the capture of ghostly voices on electronic recordings, sometimes as a form of communication in response to questions.

but it has a series of lights that light up when an electromagnetic field is detected. The more lights activated, the stronger the field. They have used this to communicate with ghosts by asking the ghost to activate the lights in answer to questions. An example would be to have it activate one light for a yes answer and more than one light for a no answer. After a series of questions, they determined that they were communicating with a woman who was waiting on a train and she was in the 1940s. The investigators concluded she must be in some kind of time loop.

That was the moment I had my epiphany. Everything fell into place, and I thought, "That is what happens. When we die, we exist in time instead of space." According to Einstein's theory of relativity, space and time are variations of the same thing. As physical beings, we exist in space-time. When we become energy beings, we exist in time-space. We transform from Homo sapiens to Homo quantumus.

CHAPTER 1

Adventures of a Fledgling Ghost Hunter

Watching Ghost Hunters inspired me to try out an investigation for myself to see what kind of evidence I could document to help prove the existence of ghosts. I knew the TAPS team had been to the old Ohio State Reformatory in Mansfield, Ohio, which is about 90 minutes from where I live. I searched online and found the reformatory's website and learned that different kinds of paranormal-related events are offered there, so I decided to check it out. My wife and I initially signed up to go on a ghost walk — a guided tour of the old prison that included information about the reformatory's history and some of the hot spots where various people had reported paranormal activity within its walls. We took my teenaged granddaughter along and didn't have high expectations of catching much in the way of evidence, but rather just taking some pictures in what was reported to be a haunted location and seeing what happened.

The reformatory was built in the latter part of the 19th century and housed prisoners until it was closed in December 1990. An operational prison still exists nearby, and photographs are prohibited through certain windows in the old reformatory for security reasons. From the moment you step into the entrance, the reformatory gives a feeling of having stepped back in time. The stone façade has a distinctly Victorian appearance, and the main entrance, warden's office and living quarters contain elegant fireplaces and woodwork that remains beautiful despite decades of disuse and that wouldn't have been out of place in an opulent mansion from the same period. The prison itself is a multi-story fortress of peeling paint and cracked concrete. Many of the dark, claustrophobic cells still contain rusting bunk frames or dust-covered books and magazines left behind by their last occupants more than 20 years ago.

I didn't have any special equipment for the ghost walk — just a couple of disposable 35 millimeter film cameras I picked up at a drugstore. We toured the prison with a guide, who told us stories about paranormal activity. It's worth noting that the reformatory also has been used as a location for several movies, including *The Shawshank Redemption* and *Air Force One*. Props from *The Shawshank Redemption* are scattered around the prison, and spots where particular scenes were filmed are marked. During the tour, we took pictures with our disposable cameras. Once I came home and got the film developed, I saw I had taken a picture of what appears to be a transparent black mass.

We had a couple of other interesting experiences on that tour that we couldn't explain. My granddaughter – a brave and adventurous soul – at one point said she felt like someone had pulled her hair when no one was there. At another point during the tour, I distinctly heard the clang of a cell door slamming as we walked through one of the old cell blocks. The tour guide had made a point at the start of telling everyone not to close the cell doors because they no longer had the keys and anyone locked inside a cell would have to stay there until a locksmith could be called to get them out. I knew it couldn't be one of our group doing closing a cell door, so it had to be something else.

Those experiences combined with the intriguing photo I had captured whet my appetite to try an actual ghost hunt, which the reformatory also offers. Their ghost hunts are overnight events in pitch darkness with participants given free rein to explore most of the prison.

Once I was bitten by the ghost hunting bug, I searched eBay and found some basic ghost hunting equipment such as an EMF detector, voice recorder and an infrared[3] camera. I didn't buy these all at once, but built my ghost hunting arsenal over time. For our first true ghost hunt at the Mansfield reformatory, my wife and I once again carried disposable 35 millimeter film cameras, but this time we also had an EMF detector. It's believed that the presence of ghosts is indicated by strong electromagnetic fields – a spike on the meter may mean a spirit is present, if other sources of electromagnetic energy can be ruled out.

We decided to try a systematic approach to documenting the possible presence of spirits in the reformatory. I carried the

---

[3] An infrared camera shoots in the infrared spectrum and captures more than a regular camera, which shoots in the visible spectrum. Photos from an infrared camera come out with a reddish or bluish hue and can pick up more than a traditional digital camera.

EMF detector as we walked through the dark building — guided only by the slivers of light from our handheld flashlights — and whenever I saw a spike or a high reading, my wife would take a picture with one of our disposable cameras. Most of our pictures didn't capture anything, but we did take one picture of what looked to me like an orb. The photo shows a single bright dot of light and doesn't look like specks of dust in the room, that can be illuminated when a camera flash fires.

Our best photo from that night was caught entirely by accident, though. There is a central area connecting the two massive cell blocks called the bullpen. As we were preparing to leave for the night, my wife noticed she had a few exposures left on the disposable camera. She decided to take a few pictures in the bullpen just to use up the film so we could get it developed. We thought nothing of it, until we got the pictures back and saw that in one of the bullpen photos there was a dark figure hovering three feet off the ground by the far wall from where we had been standing. The figure looked like it was in a wheelchair floating above a set of traffic cones sitting on the floor where some restoration work was being done. The figure appeared to be from the Victorian era.

We tried our technique of snapping a photo when the EMF reader spiked at an abandoned tunnel in Haydenville, Ohio. My granddaughter also came with us on that trip. The old tunnel once was used to transport lumber from a mill to a highway on the other side of a hill. Stories are that several men lost their lives

13

digging the tunnel through the hill. As we walked through the tunnel, my EMF gauge would spike and then again my wife would take pictures. In this case, I found it odd that my meter would spike because there was absolutely no electrical power there, just dirt and brick, so it would seem the spikes were indicating paranormal activity. Once we processed the pictures, they showed a mist forming. First one figure appears and then another, so we had captured two figures in the mist. While she was taking pictures we actually saw nothing out of the ordinary. Everything was on film.

So I was starting to capture what seemed to me like good photographic evidence of the existence of ghosts, but not every trip was so successful. We took a ghost hunting trip to a famous sanatorium in Louisville, Kentucky. This place was as creepy as you can get, and was the site where thousands of people died of tuberculosis during the decades the hospital was open. There building consists of four floor levels with a morgue on one level. There is a tunnel — nicknamed the body chute — that was used to transport the dead to awaiting vehicles so other tuberculosis patients couldn't see them.

We spent some time on each level exploring and taking pictures. I had my voice recorder and sat in different rooms trying to get any EVPs that I could. At one point as we explored the third floor of the sanatorium, we heard a loud crashing noise. They were doing some construction work on that level and apparently a wall had collapsed. It turned out it wasn't anything paranormal. Even though we spent quite a bit of time there, we really didn't catch anything on film or on my voice recorder. I know other investigators have caught paranormal activity there. I guess the spirits just didn't want to play that night.

On another occasion, I had the opportunity to visit an old mausoleum with a friend of mine. This was a unique individual who said that she could feel the presence of ghosts and otherworldly beings. I had never put much faith in this type of thing, although I am sure a limited number of people are sensitive to such things. As we walked around the outside of the mausoleum — the inside was closed to visitors — there was a double stone stairway leading to a second floor entrance. On each side of the stairs was a recessed area about thirty feet long. As we went up the stairs, we got to the top and I asked her if she felt anything. She said she didn't sense anything at the top, but had in one of the recessed areas below. We walked back down the stairs, and as we were ready to leave, she turned and took a picture of the area with a digital camera. When we came back home, she downloaded the pictures onto her computer. There, in the center of the recessed area, was a ghostly figure that looked like it was jumping out at you. It's a cool picture. I did some research on the mausoleum and it turns out a magician who actually knew Houdini was buried there in the 1930s. Folklore is that this guy is still hanging around.

Another interesting place I've visited is another abandoned railroad tunnel in southeastern Ohio. My wife and I went there with my sister. My wife stayed in the car because she wasn't feeling well, so my sister and I walked the short distance to the tunnel. At one time there was a town located there but all that is left now are a few foundations among the trees. My sister and I entered the tunnel, which is only about 50 feet long. I was about six feet in front of her when she asked if I had said anything and I said, "No. Why?" She told me that she heard a voice whisper "Hey" into her ear. I looked around and didn't see anything that would have made a sound like that. I had my EMF detector, but I didn't get any spikes. We walked to the end of the tunnel and back, taking a few pictures along the way. She had a disposable 35 millimeter camera and I had an infrared digital camera. We stood at the entrance and took a few more pictures before we decided to leave and check on my wife, who was still in the car.

When we arrived home, I plugged the media card from my infrared camera into my computer. A lot of the pictures were blurry and didn't show much at all, but I noticed one that showed the figure of a man. It was a bit blurry, but I could make out some detail. I could see a balding hairline, the outline of his nose and chin, what looked like a jacket with a collar. There is another figure to the left of him, but it shows no detail. I can tell something was there because it blocked the light from the end of the tunnel. While this was going on, we didn't see anything with our eyes, just in the eye of my camera. By the outline of the figure on the left, it looks as though it might be a woman with an old-fashioned hat with a veil. She looks like she might have a lace shawl around her shoulders. Even though we didn't see anything out of the ordinary, the figures in the picture appeared to be only a few feet away from me. My sister's pictures when developed didn't show anything ghostly.

Once I had been ghost hunting for a little while, I came across an opportunity to try an investigation in a private residence when a friend and co-worker told me about some unusual things that had happened since she moved into a house that was built in the 1960s. She told me that since she had moved in, she had heard strange noises and that the covers and pillows on a bed in a back bedroom had been removed from the bed and thrown on the floor. She would put her infant grandson in the room for naps and he wouldn't want to stay in there, and would cry the entire time he was there. She said toys would be moved and lined up on the couch in the living room. She asked if we could come over and try to find out what was going on.

My wife and I went there one evening and took pictures and tried to get some EMF readings. We found high EMF levels throughout the house and thought this might be the problem. If a location has high EMF levels, it can cause EMF sensitivity, which can result in feelings of being watched or even hallucinations of paranormal activity. I decided to spend some time alone in the back bedroom with me camera and voice recorder. I took some pictures and asked questions with the recorder turned on. There was a closet with a child's walker in front of the door. I was curious about what might be in the closet so I moved the walker and opened the door. It turned the closet only contained sheets and blankets.

After a while, we packed up our ghost-hunting goodies and went home. The next day I reviewed my photos and recordings to see what evidence I might have caught. I didn't capture anything on film, but as I listened to the voice recorder I heard a voice that shouldn't have been there. I could hear myself moving the child's walker and opening the closet door. I heard a child's voice on the recording during the time I had the door open. There was a sigh and then the word "bed." After reviewing this, the first thought that crossed my mind was that it was the spirit of a little girl who missed the baby being in the bedroom. When my friend moved into the house, there was a box of toys in the garage. One of the toys she kept was a very nice doll dressed in an antebellum dress. The neighbors had told her that the previous owners' daughter had died in the house. We decided that we would return and put the doll in the room to see if it would generate any activity. I put a voice recorder beside the doll and had camcorders on both the doll and closet door. Unfortunately, I didn't capture any more evidence. What I did get was my first EVP.

We made a return visit to the now familiar Mansfield Reformatory – this time in November. It was cold in there, but they had areas set up with some heat and they provide hot drinks to help you warm up a bit. As it turned out, the cold seemed to increase the paranormal activity compared to when we visited during warmer weather. This time, they had opened up two new areas for ghost hunting — a basement area under the solitary confinement cells and a basement area under the administration section with the warden's office.

As the night began, those of us who hadn't been there before or those who just wanted a tour went on a tour showing various areas of the prison. The tour usually starts in the cell block areas, so my wife and I went the opposite direction. The first place we went to was the area opened up under solitary confinement. It was dark, of course, but we on this trip we had green hat lights to illuminate our way. We used green lights because green preserves the human eye's natural ability to see at night, whereas a bright white light can affect how the eye processes light. I had my EMF detector and it spiked a couple of times, so my wife took pictures when that occurred. Using this technique, we caught a really spooky figure standing by a fence gate. He was pale with long, stringy hair — like a movie vampire. He looked pretty strange, but kind of cool also. We had no idea he was there until we got the picture developed.

The next area we checked out was the new area under the warden's offices. As we walked down the stairs, we could hear someone talking. We really didn't listen for what was said because we thought it was just another ghost hunter downstairs. But as we stepped into the basement, we found there were no lights on and no one was there. I really think we heard our first disembodied voices. We took some pictures but didn't catch any photographic evidence down there.

After that, we went to the infirmary area. I had my EMF and it would spike rapidly as if something was running by me. It happened a couple of times. As usual, my wife was snapping pictures and we caught what looks like an orb. It was a small ball of solid light and didn't look anything like the dust I have seen in other pictures. We went through the office areas and had occasional spikes, and we caught a definite shadow figure in one area. It wasn't a well-defined form, but more like a black mass.

After a while, we headed back down to the bullpen area — where my wife photographed a figure on a previous visit — to warm up and have some pizza the organizers had brought in. While we were at the top of the landing looking down into this area, my wife took a couple of pictures. Once again, we caught a figure by the same wall where we had caught the entity in the wheel chair on our previous trip. It looked like a thin man with short hair, and the best part was that you could see right through him. Needless to say, this night provided probably the best activity we had seen there. It just makes us want to go back again and again.

I had another opportunity for a private investigation when a longtime friend who works in an auto parts store — and whom I consider to be a very reliable person — told me about some possible activity in the store where she worked. The store was a new store that had replaced an old one that had become cramped and outdated. The new store was built on an empty lot across the street from the original location, and after the new building was completed and opened, my friend started noticing strange things happening. She and others in the store would hear noises like someone was in the back of the store when no one was there. At first they discounted the noises as mice skittering around, or maybe the building settling. There wasn't anything constant or occurring on a regular basis, just an odd sound now and then.

On one occasion, the manager was opening the store and had put cash into the cash drawers preparing for the day's business. He went into the office area to retrieve some print-outs, and when he returned to the front area he noticed that one of the cash drawers was open. He thought this was strange because he was the only one in the store and he had locked the cash drawers as was required by company policy.

On another occasion, someone was in the stock area using a scan gun to scan in the daily stock order. The person laid the gun on the stock pallet to put some stock away and when they returned the gun was missing. This person ended up finding it on the other side of the stock room in a bulk area.

On different occasions she had heard what sounded like someone walking in the back of the store when no one was there. One time she thought she heard someone walk into the office and shuffle papers around when no one was there. She asked me if I thought there could be a ghost in a new building. I told her that she would probably need to know the history of the area to determine if there could be a reason for a ghost or entity to be there. The new store is located on a busy street and there could very well have been an accident in the area in which someone was killed and still lingers at the site. There could be several possibilities why a spirit remains and it is likely we would never know. Of all the occurrences, I don't believe anyone has heard a voice or talking – just strange noises.

I had occasion to work in that store at one point and had an experience of my own. I was in the aisle where they stocked filters and heard a filter hit the floor in the next aisle. I didn't think much about it, thinking that someone just knocked one off of the shelf until the manager came running over and said he bet he knew the part number of the fallen filter. I asked him how he could possibly know that, and he told me that it was the third time in two days that same filter had hit the floor. He said each time he picked the filter up and placed it securely on the shelf. I decided that, with his permission, I would bring my infrared camera and try and catch Mr. Filter Guy.

I went in the next day and walked around taking a few pictures. I made sure there weren't any employees or customers in the area I was photographing. There was a closet and loft area, and I made sure I got pictures there because these would make good hiding places while the store was open. Upon reviewing the pictures, I noticed someone standing at the end of one of the aisles. The person had longish hair and a beard that would rule out any employee because of the company dress code. As I brought up the picture on my computer screen, it looked as though you could see through him. I had my wife print a copy and I showed it to the manager. I told him that I tried to make sure no one was there when I took the picture, but it was possible that a customer walked into frame as I was photographing. I asked him if he recognized the person in the picture as a customer or someone he knew. He was kind of startled and told me that he didn't know who that could be and that no customers would have been in the area where the picture was taken. Without my prompting him, he also pointed out the fact that you could see through part of him. I think I had caught his mysterious filter prankster.

    Having had some experiences with the paranormal and captures some photographic and recorded evidence, I next wanted to explore what it all means by journeying into humanity's past to evaluate various cultural and religious beliefs about the afterlife, and how physics might illuminate what happens after we die.

CHAPTER 2
Haunting Theories

It's a common reaction for people to be afraid when they encounter something that might be paranormal. But why be afraid of a ghost? Weren't many of these spirits once family members, friends or neighbors? Did they somehow change or become evil after their physical body died? It's my belief that they remain the same people, just without a body. They've outgrown them. Our deep-rooted fear of ghosts probably stems from lack of understanding. We need to examine our beliefs about and our fear of ghosts. There are different schools of thought about whether ghosts exists, and if they did what they are. One thought is that ghosts simply do not exist. Many believe that they can be explained by any number of natural phenomena. The noise of a house settling, the wind, or shadows caused by just about anything. They use any conceivable explanation except the idea that it might really be a ghost.

Another thought is that ghosts are products of legend and hearsay — claims that a place is haunted are solely based on superstitions and stories told to frighten people or to keep people away from a certain location.

Another explanation for ghosts is that they are hallucinations. If one or more people are in a location that is reported to be haunted, then the theory is that whoever visits this location would expect to see ghosts or paranormal activity, like a self-fulfilling prophecy. In this line of thinking, if a place is supposed to be haunted then people who go there will see ghosts when they are not there simply because they expect to.

There is a theory that ghosts are actually telepathic projections created by the living. Something along the lines of an out-of-body experience, or the manifestation of a telepathic message expressing danger, or a stressful situation such as impending doom. Although these telepathic incidents might occur, I question whether they truly could explain ghostly activity.

The most prevalent theory about ghosts is that they really are ghosts — the life-force energy of those who have departed life. Whether they are what is called an intelligent haunt or what some call a residual haunt, which we shall examine later, they are still ghosts.

The idea of ghosts has been with us throughout history. The word ghost has been used since the sixteenth century. It is derived from the word *gast*, which in old German meant a terrifying rage. A form of this word is still used today. When a person experiences a terrifying or stressful situation they are described as being aghast, which can be translated as being terrorized by an angry ghost. So, what is a ghost? The origins of the terminology used to describe ghosts has given us insight into the characterizations depicted through the ages. The ghost is created by the spirit that possesses the body and the spirit is created by the breath that ignites life. A baby in the womb is alive but many think that it becomes of spirit with its first breath and in turn a dying person leaves this world with their last breath, giving us the notion of giving up the ghost. When the spirit or ghost leaves, then the body is left behind. The spirit is the higher self that survives the death of the body.

I think the question that needs to be asked is not what creates a ghost, but what makes a ghost? Most of us in Western culture have been taught that the spirit survives the death of the body — this is a belief held by most major religions, and that the spirit or soul goes to heaven or a designated place to spend eternity. Obviously most spirits go somewhere or else every structure or house or place of residence would contain a ghost and they would be as common as having a neighbor. With as many people who have died over the millennia we would be overrun with spirits. There must be reasons or situations or some catalyst that causes a spirit or ghost to remain. I will offer a closer examination of this and offer possible explanations in further chapters, but we need to look at why some ghosts remain. There are theories that lost souls cannot find rest because of unfinished business or that they are attached to people or possessions. Legends have souls wandering eternity because of past deeds or the quest for true love. Some suggest that unmarked graves or improper or inadequate funeral rites or burials create restless spirits. Some might be there just because they want to scare the bejeepers out of someone. Whatever the reasons, some entities remain.

On the ghost hunts I have been on, I have never actually seen a ghost – yet. I know people who claim to be able to see ghosts, but I don't have that ability or sensitivity to such things. As I described previously, I knew a woman who felt a presence at a location we were at and she got a picture of what looked like a ghost to me, but I didn't feel anything. I have to rely on an EMF detector or other equipment to find ghosts or evidence of ghosts.

There are different ways in which a ghost can present itself. One way that seems to be the most common is to appear visually. These types of occurrences are usually elusive, appearing as glimpses or seeing something out of the corner of your eye. Sometimes the sighting lingers, such as when an entity moves from one room to another or disappears through a wall. Seeing such a full-bodied apparition, which is Latin for appearance or presentation, is considered to be the "Holy Grail" for paranormal investigators. Viable sightings of such apparitions are usually associated with the house or property they are in, or with items or belongings they seem to be attached to.

Some ghosts are associated with sounds rather than sight. Many complaints encountered by investigators include the sound of footsteps, knocking or banging, items being dropped, etc. Investigators have encountered footsteps close to them but were unable to capture any evidence of an entity using cameras or thermal imaging equipment. Another sound-associated aspect of a haunting is the recording of electronic voice phenomenon, or EVP. This is where a disembodied voice is recorded in a situation where it could not be heard. Questions could be asked and answers provided by a spirit voice, or the sounds of a ghost or spirit could be recorded in an empty room.

Another way a presence can make itself known is by the feeling of a presence. As I have stated, I don't have that ability but if a spirit senses that someone can, they might make themselves known this way. Another possible way a spirit can announce itself is through dreams or visions. Such instances are sometimes associated with a pending disaster or some type of unpleasantness, but such occurrences are rare and are usually dismissed.

During the nineteenth and early twentieth centuries, mediums and spiritualists would conduct séances in an effort to conjure up ghosts. They would charge a fee to contact loved ones for grieving clients and usually put on a show with banging noises and sheets floating from wires in an effort to convince unwitting patrons that they had contacted the dead. Needless to say such practices soon lost favor.

During the Victorian era, there was much interest in

ghosts. There have been stories about fairies and witches and ghosts for centuries, but as populations migrated from farming and rural areas to larger towns and cities the legends of imps and fairies that were residents of the woods and countryside were left behind. Ghosts, however, reside in both urban and rural settings. Tales of ghostly visits were very popular in the novels and stories of the 19th century — the works of Edgar Allen Poe stand out, or Charles Dickens' *A Christmas Carol*, or Henry James' famous novella *The Turning of the Screw*. References to ghosts were made in several of the novels of great Victorian novelist Thomas Hardy, and in Emily Bronte's classic novel *Wuthering Heights*, in which a ghostly visit precipitates the telling of the tale of Heathcliff and Catherine's tempestuous and doomed love. Such stories of ghosts and death helped keep the idea of ghosts in the minds of a large audience. It was clear that the Victorians seemed to have a love affair with the paranormal.

The Catholic Church has long had the concept of purgatory. This, they say, is the realm between this world and heaven, where souls go to be cleansed before entering heaven. Other religions believe that when we die we either enter heaven or hell, depending on the righteousness of the individual in question. Purgatory seemed to be the perfect place from which ghosts could come. Since they are neither in heaven or hell, nor belong to this world, they would be free to move about and do their hauntings. It was natural for the Victorians to embrace the ideas of spiritualism. The basic concept of this is the communication with the dead. The idea was that a ghost could communicate through messages in the form of knocking and rapping sounds. Mediums would fall into trances to talk with past loved ones. With the advent of photography, some would claim to have pictures of the dead that were usually double exposures, creating unusual images in otherwise normal pictures. Even Thomas Edison tried in the 1890s to create a phonograph-type machine in which to talk to the dead. One aspect of spiritualism was that anyone could participate and was not limited to any social circle or income group. This added momentum to the movement and added to the popularity. But the Victorians were not unique in their fascination with or belief that part of us lingers after death.

CHAPTER 3

A Journey into Humanity's Past

Ever since the spark that makes us human was ignited, we have had a belief in an afterlife, a continuance of our being. Somehow we have an awareness that death is not the end. In almost every culture since the dawn of history and in every corner of the world, we have had rituals to prepare us for our second existence. These rituals vary greatly from time period to time period and from one location to another, but the principle notion is the same – our spirit or soul or ghost or whatever name you want to use will exist in another place. The basic ideas from the beginning are pretty much fundamental – prepare the body for the journey, include items that the dead might use in the afterlife and develop ceremonies to help make the transition.

Funeral rites are one indicator of a cultural belief in an afterlife, and in humanity date back to one of our early ancestors — Homo Heidelbergensis. This species is believed by physical anthropologists to be ancestors of Neanderthals and Homo sapiens. The divergence from Heidelbergensis into Neanderthals probably occurred around 300,000 years ago in Europe and into Homo sapiens about 200,000 to 100,000 years ago in Africa. The Neanderthals retained most of the features of Heidelbergensis such as large brow ridges, protruding faces and lack of prominent chin. The first Heidelbergensis fossils were found Oct. 21, 1907, when worker Daniel Hartmann found a jawbone in a sand pit. Heidelbergensis is credited with having a large brain case with a typical volume of 1,100 to 1,400 centimeters, compared to the 1,350-centimeter average of modern humans. The dimensions of the outer and middle ear suggests that they had hearing sensitivity close to modern humans. They probably were able to differentiate many different sounds. Heidelbergensis is thought to be the first of our ancestors to speak, and probably had an early form of language.

Recent findings suggest that Heidelbergensis may have been the first human ancestor to bury their dead. In 1997, a team located more than 5,500 bones dating to at least 350,000 years old in the Sima de los Huesos site in northern Spain. The site contained fossil remains of perhaps 28 individuals along with the bones of various animals. Along with the fossil remains was a red quartzite ax that they dubbed Excalibur. It seemed that the stone ax was a kind of ritual offering to be used in the next life. The offering of such an item would be significant because something like an ax was used for their very survival.

The oldest modern human fossils were found along the Omo River in Ethiopia. The fossils found on either side of the river were named Omo 1 and Omo 2. These remains were dated using new dating techniques at 195,000 years old. It should be noted that the Omo 2 remains are considered much more primitive than Omo 1. This suggests that modern humans and less than modern humans existed at the same time in the same areas. Many scientists think that what makes us human is our modern behavior initiated by our modern brains. Modern behavior starts to show itself sporadically around 70,000 to 80,000 years ago, but doesn't take hold until around 50,000 years ago. This event, sometimes called the "Great Leap Forward," occurred at the beginning of the Upper Paleolithic era, or the early Stone Age.

I personally believe that when modern humans first appeared they were as intelligent as we are today. Technology evolved as needs arose and populations increased. A few bands of hunter-gatherers didn't need much more than the means to take prey and cook meals, and temporary shelter from the elements. As populations began to grow, we needed more and better resources to survive. The human population seems to have been reduced to about 2,000 individuals somewhere around 70,000 years ago, which coincides with the worst part of the last Ice Age. This suggests that humans at that time had to be intelligent enough to survive and this enabled them to establish themselves far from Africa. Homo sapiens may not have been common until around 50,000 to 60,000 years ago, and this could be taken as the real origin of modern humans.

If we start from the basis that when modern humans first

appeared that they had a fairly high degree of intelligence, and that as far back as Heidelbergensis we had some form of ritual burial which indicates a belief in some form of afterlife or extended existence, we should look for examples that we had some forms of artistic expressions to show that there was more to our existence than mere survival.

In the Cave of Pigeons in Taforalt, located in eastern Morocco, were found small perforated sea shells that have been dated at 82,000 years old. The find consists of 13 shells of the Nassarius gibbosulus species. The shells were intentionally perforated and some still had traces of red ocher. The dig, led by Abdeijalil Bouzouggar from the National Institute of Archeological and Heritage Sciences (INSAP, Morroco) and Nick Barton of the University of Oxford in the United Kingdom, had been conducting a study of the site for about five years. They paid particular attention to studying the shells and have determined the shells were gathered when they were dead along the beaches of Morocco, which were located about 25 miles from the caves. They have determined that the shells were gathered, made into adornments and colored red for symbolic uses. These shells also showed traces of wear, which mean that they were used for a long time. They also were very similar to those found in Paleolithic sites at Skuhl in Israel and at Oued Djebbana in Algeria. This shows that about 80,000 years ago, the inhabitants of the Mediterranean area had the same types of artistic expressions and symbolism, which shows a degree of intelligence and artistic aptitude.

Music is another form of artistic expression that dates back millennia. In a cave near Ulm, Germany, a flute carved from a griffon vulture bone has been unearthed. The bone was hollowed and five spaced holes were drilled into it to form a crude musical instrument that has been dated to about 35,000 years ago. All of this can be tied to the notion that we used symbolism, artistic expression and our basic intelligence to understand the complexities of our existence, which includes the beginnings of a belief in an afterlife.

Some of the earliest known burial sites belong to the Neanderthal. Many sites have been located in Europe and the Near East. The way the remains were placed indicate ritualistic elements, as they were found in fetal or sleeping positions. Some have flowers or plants placed in their hands or on their bodies. They also used red ocher in symbolic ways. Some were interred together as family or tribal units remaining together in death. In some locations, food, stone tools and various items were placed with the bodies, implying that Neanderthals believed in an afterlife and that the items left would be needed. Some of the plants left in the graves had medical properties that were used for specific purposes. Some of the remains have been dated to about 100,000 years ago.

In 1969 in the Lake Mungo region of western New South Wales, Australia, the remains of what is known as the Mungo

Lady were discovered. The interesting part of this find is the fact that her bones were burned, crushed and buries — making Mungo Lady the earliest human cremation ever found, dated to about 42,000 years ago. This reflected a change in burial practices. Another skeleton was discovered nearby about five years later. This find is known as Mungo Man, also dated to about 42,000 years ago. He was found buried with his hands crossed in his lap and sprinkled with red ochre, and is considered the earliest known ritual burial in Australia.

In Austria, scientists have unearthed the remains of three infants. The first of this type found in Europe dates from the Upper Paleolithic period around 27,000 years ago. The remains were covered with a mammoth shoulder bone held up by part of a tusk. One of the infants was adorned with more than 30 ivory beads. This shows that even children were honored on their way to the afterlife.

When considering ritual burial, precious adornments and artistic creativity to gauge early human intelligence, one of the most outstanding creations has to be ancient cave art. When you consider this as a non-necessary task, the sophistication and artistic depiction is amazing. These pictographs date to some 30,000 to 40,000 years ago. It was originally thought that the pictures were printed as a form of "hunting magic" by which these animals would magically appear just by painting them, thus they would have plenty to eat. I think that the originations of these theories didn't give much credit to the creativity or intelligence of our ancestors.

Archaeostronomer Chantal Jègues-Wolkiewiez in some amazing work has uncovered some very interesting aspects of prehistoric cave paintings. In taking measurements of 130 caves in the south of France, she discovered that all are oriented in the direction of solar events – sunrise and sunset at summer and winter solstice and spring and autumn equinox. The paintings also showed seasonal variations in some animals depicted — some showed summer or winter coats depending on the times of the year that the sun brightened the caves. The paintings also showed when various animals were in mating season phases. On a small bone carved 32,000 years ago at Abri Blanchard in Dordogne, France, were 69 engraved incisions that calculate the moon's position over a 69-day period. This would indicate that the maker had astronomical knowledge in Paleolithic times. All of this should be taken into account as we consider how intelligent our ancestors were and how they perceived the world, the heavens and the beyond.

As human civilizations progressed through the ages, the beliefs and traditions of an afterlife progressed with them. If we look at the Etruscans living in western Italy before the Romans, they considered death a celebration and a continuation of their lives. They had cities of the dead – necropolis – located in the hills. Each tomb copied Etruscan houses, and it is from such tombs that we have been able to learn about aspects of their everyday lives.

In ancient Greece, the tombs and rituals for the wealthy were very extravagant. They used gold and jewels as offerings — and the more wealth someone held, the more elaborate the tomb. In early Greece, single graves were dug in the ground or cut out of rock. As time progressed, they became more sophisticated with multiple graves rising, raised mounds, masonry tombs, or underground chambers. Elaborate burial rituals went hand-in-hand with the sophistication of the tombs. The ritual first involved laying out the body. Women would wash, dress and cover the body with flowers. The mouth and eyes were shut to prevent the soul from escaping. The corpse was laid out for viewing for two days, and mourners would wear black in honor of the departed. The women would position themselves at the head of the place where the body lay and the men would stand with their palms out to the gods. When it was time for the burial, which came before the dawn of the third day, the corpse was taken by cart to the tomb. The men led and the women followed. At the funeral, the body would be placed in the tomb along with pottery, jewels or other personal property.

The Greeks believed that after death, the soul would leave the body to enter Hades — the realm of the dead. The soul could be seen, but not touched. If the body did not receive a proper burial then the soul would be trapped between the world of the living and the underworld. Legends exist of ghosts appearing to the living to ask for proper burial. Apulcius provides an example of a type of someone being haunted in a dream. In "The Golden Ass," a baker is murdered by a ghost his wife conjured to kill him because he wanted to divorce her. His spirit visits his daughter in a dream to tell her what caused his death. Desecrating a tomb also could lead to being haunted by a spirit.

The ancient Greek belief in spirits led to annual festivals. The feast of Anthesteria was held sometime in February or March in Athens. During these events, everything stopped. Businesses and temples were closed. The doors of houses were covered with pitch and hawthorn leaves, and every family member made an offering to the dead. All of this shows that the Greeks were fascinated with the afterlife.

Ancient Egyptians held very strong beliefs in the afterlife that were integral to their culture. Osiris was their god of the underworld, and their legends tell that Osiris was killed by the god Seth. Osiris' body was torn apart and pieces flung throughout Egypt. The goddess Isis and her sister, Nephthys, took the pieces and gave a new life to Osiris, who then became ruler of the underworld. The Egyptians thought that each individual had two souls. One, called the *ka*, remained with the body throughout life. After death, the *ka* left the body and entered the realm of the dead. They thought the *ka* could not exist without the body, so they preserved the body through mummification. The other soul was called the *ba*. This was believed to be the soul that actually animated the person, and was depicted in ancient Egyptian art as a bird with a human head.

Various digs have uncovered specific practices concerning these strong beliefs in the afterlife. Many jars containing mummified organs for use in the afterlife have been found. The pyramids — massive tombs for dead Egyptian rulers — each had an inner chamber that contained the mummified remains, servants of the departed, and items that were to be used in the next existence.

Around 1,000 BCE, the prophet Zoroaster formed a religion in ancient Persia that was influenced by ideas of good versus evil. The god of light was involved in a struggle with the god of darkness, who dwelled in the lower world. Individuals were to support the forces of light and were judged according to their good or evil deeds. In the afterlife, Zoroastrianism states that for three days after death the soul stays at the head of the body. All of the good and bad deeds are put into a ledger with evil being counted as debits and good actions being rated as credits. The soul then takes a journey to judgment on the Chinvat Bridge. At the center of the bridge is a sword. Below the bridge is hell. The soul is taken to where the sword stands. If the soul is considered good, then the sword shows its broad side. If the soul is evil, the sword stands edgewise and does not grant passage. The soul is cast into hell.

As we have seen, beliefs in an afterlife have been with us throughout history and have covered many geographical areas. As an intelligent species, we have always been curious about what lies beyond our all-too brief residence in our physical bodies.

When looking at Viking and old Norse traditions, we see that as in other traditions a person's afterlife was determined by how they lived their life. They believed that if the departed had an unexceptional life, they would travel to a place called "Hel," which is where the English word "hell" is derived. This place lies to the north and is underground. They thought of this place as being cold and damp, and the spirits dwelling there existed in a dreamlike state. It was not an exceptionally pleasant place, but it wasn't torturous either. The souls in this realm were considered to be in a long sleep.

There was thought to be another region beneath Hel where evil or bad souls went. The name of this place was Nastrond, or the shore of corpses, where a serpent called Nidhogg would gnaw at the souls of the dead. It was located on the shore of an ice cold underground sea and the poor souls would sleep in a hall made of snakes that dripped poison. It looks like it was a bad idea to be an evil Viking.

Those who lived good and righteous lives would travel to Asgard, the home of the gods. This is the place where they could spend eternity in happiness. Heroes who died in battle would go to Valhalla, the hall of the slain. At this place, they would live with the god Odin and would spend the day fighting with each other. At the end of the evening, they would arise healed of their wounds and spend the rest of the night feasting.

When we look at superstitions and traditions concerning ghosts in different cultures, then the Chinese certainly have the most intricate among anyone. Confucius once said, "Respect ghosts and gods, but keep away from them." According to Chinese folklore, ghosts can take on many forms according to the way in which they died. The belief in ghosts are part of the Chinese tradition of ancestor worship, and many Chinese use mediums to contact the spirits of their ancestors seeking guidance for themselves and their descendants. In traditional Chinese Mythology, Yan Hang, also known as Yanluo, is the god of death and the underworld. He is the one said to judge the dead and determine the events of their afterlife. The Chinese word for ghost is *gui*, which is combined with other symbols to describe the different categories of spirits or ghosts. Here are examples of the different types of *gui*. It seems that if you were in a haunted Chinese cemetery, you would probably need a score card.

- *Diao Si Gui*: the ghost of someone who has been hanged.
- *You Hun Ye Gui*: the wandering ghost of someone who has died while traveling or being far from home and their soul or spirit has not returned home.
- *Gui Po*: the spirits of servants or helpers that often take

the form of a kindly old woman and return to continue serving the living.
- *Nu Gui*: the spirit of a woman seeking vengeance after committing suicide because of being wronged or abused.
- *Yuan Gui*: a ghost seeking justice after dying a wrongful death.
- *Shui Gui*: the ghost of someone who has drowned and still inhabits their watery grave, pulling victims under and taking over their bodies.
- *Wu Tou Gui*: a wandering headless apparition.
- *E Gui*: the ghost of someone who was consumed by greed and is sentenced to go through eternity suffering hunger and want.

As we can see, in Chinese mythology the way a person dies wields more influence over the way the soul spends eternity than the way the person lived, which is contrary to Western beliefs. In Christianity and other monotheistic religions, the concept of death is that the soul (singular) leaves the body to enter some form of afterlife depending on customs associated with a particular religion. According to the Chinese and their basic theology of ancestor worship, the soul of a deceased person consists of two parts: the *po* and the *hun*. When death occurs, these factions split into three parts. The *po* stays with the body and remains in the grave; another goes to judgment, and I believe Yan Wang would determine which *gui* would be appropriate; and the *hun* would be housed in the ancestral tablet. The *po* and *hun* are not considered immortal and need to be maintained through offerings. Eventually the *po* and *hun* end up going to heaven, with the *hun* arriving first.

The use of mediums is commonplace and is closely associated with ancestor worship. The medium contacts those on the other side and asks the spirits what they might require. These needs or wants are then met through the burning of offerings. In exchange for their offerings, the living are rewarded with different degrees of prosperity. In yet another tradition, it is important that the correct rites are performed when someone dies because the body still has the ability to affect the living through the *po*.

The Chinese have a centuries-old tradition known as the Ghost Festival. This festival occurs on the fifteenth day of the seventh month of the lunar calendar. On this day, it is believed that the deceased visit the living. Descendants pay homage to their ancestors by preparing food offerings, burning incense and offering sacrifices of newly harvested grain in the name of the departed. During this festival, people refrain from getting married or moving into a new house, especially at night because a wandering ghost – *You Hun Ye Gui* – might attach itself to them and cause bad luck. They also refrain from going near water out of fear of a *Shui Gui*, or ghost who has drowned and wants to take over their body.

In the Americas, the Native Americans had varied beliefs in survival after death. Most traditions believe that we have more than one soul — a free soul that separates itself from the body when we die, but still keeps its individuality, and another that animates the body and does not survive death. Beliefs about where the soul goes after death vary greatly from area to area. In some cultures the otherworld is a mirror of the current existence, which probably gave rise to what some call the "happy hunting ground." Many traditions of the Plains tribes see the departed as living on a rolling prairie, living in teepees and hunting abundant buffalo. Some of the other Native American cultures see the afterlife as a dark, gloomy state similar to Hades. Peoples such as the Inuit believe in reincarnation after death.

A lot of Native cultures had individuals known as shamans. They were seen as healers and guides for the souls to find their home in the afterlife. They also were mediators between the tribe or community and the spirit world. Shamans communicated with helper spirits, which usually took the form of animals or forces of nature. They also talked to the spirits of the dead themselves.

The idea that the newly dead must make a perilous journey to find their place in the land of the dead is common in both North America and South America, and emphasizes the shamanic practice of using special rituals to guide them to the otherworld. When the departed could not find their way or wanted to remain with their family – causing a haunting – a shaman was brought in. After achieving a state of trance, the shaman would convince the spirit to leave the family alone and takes the soul to the realm of the dead. Usually the Milky Way is seen as the road the spirits take on their journey.

To the Aztecs, there were four different realms of the dead that correspond with the four directions. If the dead were warriors who died in battle or were sacrifice victims, they went to the eastern paradise and were companions of the sun. Women who died during childbirth went to the west and also became kin to the sun. Those who died by drowning or events caused by water or rain went to the southern paradise, which was an area free of sorrow and had a lush, tropical garden. The other realm to the north was for all the others who had died from other causes. This area was considered not as pleasant as the other ones.

If we consider the existence of beliefs about the afterlife since the beginning of human existence, we must then ask why this tradition is so pervasive across cultures. We must ponder our place in the whole of the universe. What are we doing here? What is the purpose of our existence? We have always been searching for reason why we are here. The known facts are clear. We began, we have an existence, and we continue. But as intelligent and inquisitive creatures, we try to reason our place in the immense universe.

When we seek answers about the origins of our existence,

our first thought is to our parents. We ask why it is necessary for us to be here, or was it necessary for our parents to be here, and so forth and so on. We find ourselves asking why there are human beings at all. Why is there an earth to live on? Why is there a universe with its vast distances? As we ask these questions, we assume that there are answers to the questions at hand. We might not know the answers, but we are sure there is an explanation if we can only find it.

There is a general principle at work here — the principle that for everything in existence, for each event that occurs, there is something that necessitates its existence or occurrence. This principle is sometimes referred to as the "principle of sufficient reason." So when we start to ask questions about the existence of life, death and the universe, this principal leads us to assume there must be an explanation. The universe and everything in it might not have existed at one point, so therefore there must be a reason why it does. The universe could not exist uncaused. It should be incorrect to believe that the universe exists because of its own nature to.

So the question is: Does everything have a cause? It seems as though it does. Even when we don't know what the cause is, there is sure to be one. Even if we can't find the cause of an effect, our failure to have a ready answer could simply stem from lack of knowledge. We think that if we learn more about the causes and the physical laws that govern them, we can find explanations for all types of phenomena.

We also must assume that the "principle of sufficient reason" not only applies to this world or existence, but to all worlds or existences. It can be said that this is the basis of rationality. There are conflicting views of this principle because some argue that strange occurrences in quantum mechanics have no explanations or natural causes. This type of circumstance brings us into the realm of mystery. Does the question of being express a mystery? Is this a question for which there is no answer? Or is the answer we seek the ultimate explanation that is the most profound and far-reaching explanation? The question presupposes that there is an explanation of why everything exists.

If we narrow this down to living creatures, it seems that the reason for a species to exist is to reproduce. Considering that a vast majority of all species that have ever lived are extinct, one must ask if this is the prime reason for being. The failure to evolve or adapt to changing circumstances dooms them to oblivion. There should be a higher priority to life than eating, sleeping, and making babies.

Most species that live in any particular eco-system tend to have a balancing effect. There doesn't seem to be any over-abundance of predator to pray unless the balance is changed by outside influence, such as man. If we examine the perceived reasons for human existence, the progression is to learn what we can in school, get married, raise a family, and contribute in a positive way to society before retiring and ultimately facing death, which is where we receive our rewards. We do not go through this process of existence blindly. We often question the direction of our lives. We have the ability to look above our everyday routines and ponder the unknowns of the universe. The instinctive ideas about an afterlife which were instilled in us at a very early time influence us greatly. It is the mystery of the otherworld that has influenced cultures from the beginning. The ability to imagine what the next realm will be like — because the truth about such a realm is beyond our ability to perceive it — has controlled culture for thousands of years.

CHAPTER 4
God and Ghosts

When we think of an afterlife or the existence of a heavenly place or the possibility of hell, the thought of a supreme being or god has to be taken into account. The question is "Does God exist?" For many cultures over the centuries, questioning the existence of god would be considered blasphemy. Our stories of creation have a divine being bringing everything into existence. Every religion throughout history and in every corner of the world has taken for granted that a god or gods exist. We are supposed to accept this existence simply on faith or the authority of someone who is said to be influenced by or in contact with the divine.

There are many ideas concerning the existence of god. In some cultures the idea of the nature of god is personal, as are the teachings of Christianity, Judaism and Islam. Then there are those who think of god as non-personal, such as some Hindu sects and in Theravada Buddhism. Some see god as a ruler of the world and a being that influences history, while others such as Neo-Platonists and Stoics deny this. According to some religions, the divine being becomes incarnate in the world and to others it does not. We seem to have a mass of conflicting ideas about god.

If asked if you believe in god, the answer might be "Which one?" or "Which way?" Theists in one religion might be considered atheists in another.

It is difficult to understand that your religious beliefs or practices are determined by where you are born, but the teachings of mathematics and science are all the same. To some, this suggests that there is no truth to the matter of the existence of god and the long history of conflicts between various religious factions have all been in vain. Another response to this would be that with today's world of rapid communication and interaction, diverse religious factions learn from each other and are seen to be gradually converging. It could also be said that this history of differences and conflicts could be explained by different perceptions of the same reality. Furthermore, there is the possibility that the truth of some religious claims can be attributed to ordinary and everyday events involving the forces of nature. Where some claims can be attributed to natural events, there could be underlying events which lead to the existence of a divine reality that determines religious truths.

Take Christianity, for example. The idea that Christ was uniquely divine and the only son of God presented him as the sole mediator between God and man. This would mean that their claims are true and those of non-Christian faiths were false. Since the evidence provided involves the religious experiences of followers within the religion, their truth seems absolute. If we look at the broader picture, any divine creation which was sent by God to offer salvation to man would encompass all men regardless. One must assume that all of the monotheistic religions of the world — no matter the name they use — worship the same god. Because of ignorance or cultural or geographic isolation, they insulate themselves from other religions.

So is the thought of a god logically coherent? We must ponder the problem that if god is all-powerful or omnipresent there is nothing that he (or she) cannot do. But there are things that cannot be done — seemingly impossible things like putting an ocean on top of a mountain, or having everlasting peace on Earth. One explanation would be that whatever God does is not a contradiction of terms and that our inability to perform an illogical task is not a limitation of power. As St. Thomas Aquinas said, "Nothing which implies contradiction falls under the omnipotence of God."

Another aspect of God is said to be self-existence. It is of God's essence to exist. God exists by a necessity of God's own nature. And as God's nature, one would consider that it would be infinite and unlimited. We understand this as a matter of faith.

When we take the Biblical account of how God created the world and all the creatures in it there is a conflict with current scientific theory. If we take this account literally, it discounts Darwin's evolutionary hypothesis. The evidence for the Theory of Evolution and the current age estimates for planet Earth are overwhelming, which means the Biblical account cannot be literally true. What the Biblical account actually contains is the essential truth that God is the creator even though it is not expressed in a scientifically provable manner. We need to distinguish between stories and divinely revealed truth even though they are expressed incorrectly. We need to accept this on faith. In order to accept this for any reason, we must look for any kind of evidence to justify a belief rationally. The evidence itself must be rationally justified. If evidence is lacking to come to a firm conclusion we must accept something without rational justification. We must accept on faith. When we make an argument for the existence of God, we need to use information provided by scientific observation and observations based on faith. Since the universe exists and everything in it has a cause or effect, it must have been created by God. In some way or another, this argument has been written by philosophers and theologians for millennia. There is a belief that because the universe might not have existed, there must be a reason why it does.

Another argument for the existence of God is the theory of intelligent design. This goes on to say that with the complexity of plants and animals and their ability to adapt to their surroundings and one another, there must be an intelligence involved. A case can also be made that the natural selection process is equally efficient at making adaptations, but the question of complexity still remains. When we do consider intelligent design we need to ask, "Who is the designer? Who designed the designer?" This must take into account God as the creator. God would have to be the necessary being whose existence or nature would be self-explanatory. The main idea is that a logically necessary being is contained in the essence of nature.

When we examine the Biblical conception of God, we see a perception of the supreme being. As Anselm stated in his ontological argument, God is a being "than which nothing greater can be conceived." By definition, God is absolute perfection. It is argued that a being that did not exist would be less perfect than one that does. If we imagined gods like the ones worshiped by the ancient Greeks and Romans, there would be flaws simply because we can't imagine such perfection. Since God is a being which is greater than can be conceived, then there must be a real existence.

So how do ghosts fit into a framework of religious traditions? When we think about ghosts or hauntings, we assume — logically — that a ghost is the spirit or the remaining life force of a human being. From what we have been taught throughout history regarding afterlife, this would be a foregone conclusion, yet when we add into the equation the teachings of the Christian faith then things get a little messy. We know that the Catholics have the idea of purgatory, where ghosts could actually exist, but most other Western religions do not accommodate this.

According to the Bible one-third of the angels rebelled

against God. The leader was Lucifer, who, along with his army of rebel angels, was defeated and banished from Heaven. Lucifer has a great dislike for man because he was created in God's image and was jealous of man's relationship with God. His goal is to deceive and fool men into rejecting God, and he developed a plan for the demise of humankind. He would deceive some by showing them ghosts, while others would believe in evolution and shun the creations of God. There are some who would worship pagan gods or believe in no god at all. We have those who worship greed and power sacrificing everything in the name of profit. To me, this is the mortal sin. Lucifer or Satan sends fallen angels to deceive and create doubt within mankind. He would like man to share in his misery and punishment from God.

As stated, one of the deceptions that Satan uses is the appearance of ghosts. These are presented for us to believe that they are the spirits or souls of people we have known or cared about who have died. They have pushed back the curtains of death and returned to their loved ones. This is in direct contradiction to what the Bible says what happens when we pass on. The Bible says that the righteous shall enter the kingdom of Heaven and the wicked shall spend eternity laboring in Hell. But if the ghosts of our loved ones continue to exist on Earth, then can't be true that every soul goes either to Heaven or Hell after death. Christians would say Satan has succeeded in casting doubt about the existence of Heaven and Hell, with the idea that humanity then would question everything else in the Bible, because if this isn't true then what else is untrue?

The Bible tells us that in truth these ghosts are not really souls or spirits of men and women we knew and loved, but these are fallen angels pretending to be people we have cared about. Some so-called ghosts have appeared to some, saying that they want to help them and want to provide them with profound knowledge or offer them unrealistic comfort. This can be quite convincing to the person involved. The person or persons involved don't seem to understand that their loved ones more than likely wouldn't possess such knowledge or information. Such fallen angels, having been around since before mankind was created, would have this knowledge. Other ghosts have a more malevolent nature and want to either scare or harm someone. I have been in several haunted locations and I have never encountered a ghost or spirit that did anything like that. I am not aware of any investigators who have experienced an event such as this.

I understand the position the church needs to take to justify the foundations of Christian faith. But I believe we must take into account the context under which this was written. These ideas were written thousands of years ago. They were trying to offer an explanation for something they had little information about. This is not an attempt to say they were either right or wrong, but they were addressing a situation with limited knowledge. Even with divine inspiration, the knowledge of man could be limited.

As our information about ghosts and past lives increases then, like anything else, we will have a clearer picture as to what is actually happening. This is not an untruth, just limited truth. As with anything written in the Bible, the reference to such events could be interpreted as not addressing all ghosts or ghosts in general, but could be read as parables about what might occur if faced with the specific situation. If someone was approached by the supposed ghost of a loved one in the described manners, then we are told by the Bible that they are not the spirits we knew but fallen angels in disguise. The nature of true ghosts isn't explained by this particular situation, and I believe that even if we dwell in the kingdom of Heaven, we could still have visitation rights to return as spirits.

The Jewish faith allows for a different approach to ghosts than does Christian beliefs. The Hebrew word for ghost is *ovoth*, but there also is another word used — *dybbuk*. The *dybbuk* is an entity that can control a living body through possession. The actual root of the word is to cleave or to cling to. In the Christian faith, to be possessed by an entity is considered to be a very bad thing since this is associated with demonic possession and Satanic forces. In Judaism such possession is not considered evil. One circumstance under which possession by a *dybbuk* might occur is if a spirit has unfinished business and seeks out someone who has a similar situation. The spirit can attach itself to the host in an effort to find guidance or a solution to the problem. It is quite possible that the person may not be aware of the possession at all.

There also is a belief in the Jewish tradition that there exist spirit guides who assist earthly spirits, but only do so for a short time. There are examples of ghostly possessions in the Old Testament. Samuel 18:10 describes a time when King Saul is taunted by an evil spirit and Kings 22:20-23 tells of prophets possessed by the spirit of someone who tries to trick the king into war. There also exists in Judaism a ritual for exorcism. The ritual calls for a rabbi to blow a horn called a shofar, and it is thought that the sound emanating from the horn would drive the ghost from the host body. The rabbis would actually try to communicate with the ghost in an effort to try to help the spirit with whatever business or circumstances created the possession in the first place. Contrast this with Christian rite of exorcism, where the spirit or demon possessing a person is banished to hell.

In 1947, a group of manuscripts were discovered hidden in a series of caves near the Dead Sea that came to be known as the Dead Sea Scrolls and have been associated with an ancient Jewish sect called the Essenes, although in very recent years some scholars have questioned that association. The Essenes were a sect that established a monastery on the shores of the Dead Sea near Qumran around the middle of the second century BCE. They lived there until the Roman-Jewish war of 66-70 CE.

In contrast to some teachings of Judaism and early Christianity, the Essenes believed in an immortal soul. As written by Josephus in *The Jewish War*:

> It is indeed their unshakeable conviction that bodies are corruptible and the material composing them impermanent, whereas souls remain immortal forever. Coming forth from the most rarefied either, they are trapped in the prison house of the body as if drawn down by one of nature's spells; but once freed from the bonds of the flesh, as if released after years of slavery, they rejoice and soar aloft. Teaching the same doctrine as the sons of Greece, they declare that for the good souls there waits a home beyond the ocean, a place troubled by neither rain nor snow nor beast, but refreshed by the zephyr that blows ever gentle from the ocean. Bad souls they consign to a darksome, stormy abyss, full of punishments that know no end.

The thoughts and writings of the Essenes, as well as the teachings of early Jews and Christians, helped form the historic concepts of the standard Christian belief in everlasting life.

When we turn to Islam, there are no specific teachings about the existence of ghosts or spirits. It is thought that the spirit of an individual remains with the body in the grave until the time of judgment. Islam does have a tradition concerning *zars*, or spirits that possess someone — usually women — causing sickness and irrational behavior.

Some cultures where Islam has become the dominant religion also have stories of *jinns* that are thought to be entities that were created before humankind and live in the. Some of these could be good and some are evil. The *jinn* or *jinni* is where the word "genie" comes from. The traditional idea is that the concept of ghosts came from a time before Islam and is only superstition. Before the time of Muhammad, the *jinn* were a vital part of Arabian culture. They were said to have influence in human affairs — both good and bad. Apparently they had the ability to appear as wild animals. Muhammad had the impression that *jinn* and humans were separate species but were all under Allah. This would put the *jinn* in a similar category as angels and demons in Christianity, but angels in Islam were in a different category from *jinn* and humans.

In some segments of Islam, the *jinn* are separated into two groups, the first being an evil version that did not originate from humans, and the second that did originate from humankind and could be seen as a ghost or specter of a person after death. These *jinns* are said to be bad because there is really no reason for such a ghost to exist. It is thought that someone who is murdered or meets a nasty end generates a negative *jinn*. In the meantime, all others wait in their graves for judgment.

Where we consider the properties of ghosts and spirits the common or most natural foundation from which to offer explanations or reasons is usually one of spiritualism or religion.

As part of any discussion concerning such matters it should include the views of atheists or those who profess a rational explanation for creation and the existence of things. There is one difference between what is considered supernatural and what is considered paranormal. Most of us think that either word describes the same thing and that we could use either word interchangeably to describe an occurrence. The supernatural is seen as a disruption of natural occurrences and such disruptions are caused or precipitated by entities or apparitions. The paranormal on the other hand consists of occurrences that seem to be beyond the explanation of our current views or understanding of the laws of physics. In the supernatural world, an entity causes disruptions by means of banging or noises, disembodied voices whether heard or recorded, visible or recorded sightings and the movement of items from one place to another.

The supernatural also requires a basis of faith. Faith offers an explanation that these entities are a product of God or an everlasting soul (such as from purgatory) or some other magical source and are believed to be above or beyond nature. This concept of the supernatural is a view not shared by atheists, needless to say. This leaves the paranormal. The idea of the paranormal is that nature is not being represented as the way it should be. Paranormal means "beyond the normal". As we know, the hypotheses of science and what we understand about the universe are in constant flux. What might be considered paranormal becomes the normal. One example would be quantum entanglement, which we will review in later chapters, or the works of Albert Einstein, who changed views of the natural world dramatically.

So, how does the possible existence of ghosts fit into atheistic thought? The idea of a soul or spirit might not be specifically supernatural but rather paranormal — something we simply haven't explained yet with science. Atheists reject concepts of the supernatural because they reject ideas based on faith. Scientific reasoning has long held that there is no verifiable evidence of the existence of a soul or spirit. But with no actual scientific study of the question — no formation of a hypothesis, no testing — there's no foundation of empirical knowledge to draw upon to confirm such a conclusion. The question might be asked but the ability to answer the question has never been established. In some ways, declaring that ghosts cannot exist also is a leap of faith.

An atheist might say that our understanding of consciousness does not merit the use of resources to pursue a question that they think doesn't need to be asked at all. Nevertheless, with the increasing popularity of ghost investigations and the evidence that is presented, there stands the real possibility that ghostly phenomena are genuine. Cliff Walker of Positive Atheism Magazine drew a distinction between the supernatural and the paranormal and suggested in an article that the paranormal isn't incompatible with atheistic thought. Perhaps addressing these phenomenon from a framework of the paranormal rather than the supernatural would allow consideration of the possibility that such entities exist — and allow for legitimate lines of inquiry and evidence-collecting. It is entirely possible that the human life force can continue and this I believe this continuation is entirely natural and explainable within the laws of physics, and that we just need to continue collecting evidence to establish the existence of these phenomena. It could be the nature of humans for their intelligence or life energy to continue after the biological body dies as what I have dubbed Homo quantumus, or energy beings that exist in the dimension of time. Being explainable with physics, there's no need to rely on beliefs in religion or the supernatural as the source of these entities' existence. I realize atheists may continue to scoff at the idea, but I think describing the phenomena in the context of known laws, theories and hypotheses opens the door for rational discussion.

Regardless of the culture or their place in time or location on the planet, most cultures have traditions of an afterlife, and along with this a belief in ghosts. We might categorize a spirit as a soul that has transcended the earthly plane to heaven or hell or Valhalla or the underworld depending on the tradition you choose, and a ghost as a spirit that is earthbound for any given reason. For descriptive purposes we bunch all of these words together to reference the same thing. Since the first ritualistic burial we have recognized a separation of body and soul. Our life-force continues and we have accommodated stories and rites and pageantry in an effort to make some kind of sense of this.

Statistically speaking, even in this day and age of technology and science, there remains a pervasive underlying belief in the paranormal. The popularity of ghost investigation shows on cable is a growing phenomenon. About a third of people believe in ghosts. Twenty-three percent claim to have either seen or felt the presence of a ghost. Nineteen percent believe in the existence of spells or witchcraft. Forty-eight percent say they believe in extrasensory perception. The people most likely to say they've had a paranormal experience involving a ghost are single or Catholics, or those who are not regularly active in church. Politically speaking, thirty-one percent of liberals could see a ghost as opposed to eighteen percent of conservatives. Three people in ten have woken up sensing a presence in the room, and the ones sensing this usually are single. Fourteen percent – usually men and those below a certain income level – claim to have seen a UFO. Twenty percent say they are superstitious. With all of this and the ever-increasing advancement of knowledge, we still have an underlying sense that there is something more to this than we can account for. The mechanism from which we created all of our rituals to explain the unexplainable still exists in all of us. As for ghosts, even if there are no ghosts the words that I am writing will outlast my body. This is the sum of my intelligence and therefore my ghost, which will last as long as anyone reads this. There can be other types of ghosts than the ones that go bump in the night. In leaving a legacy, one leaves a ghost.

CHAPTER 5

Confronting Death

Once we have established the possibility and probability of there being a God, we should examine the ramifications. We need to look at the meaningfulness of existence because this is how existence is valued. We can also say that values and morality are based on reason. The basic message taught by religious leaders throughout history has been based on reason and morality except for when the message has been twisted to control the masses. It can be said that immoral acts are contrary to reason and are irrational because they are inconsistent with the basic principle of rationality. We must act appropriately and establish rules and laws that reflect rationality. When God created the world, God determined that certain issues and policies were morally right, while others were considered wrong. The rightness and wrongness are not based on what we consider to be right or wrong, but the objective moral values are ultimately decided by God.

We have long ago established the core values that have determined morality – consider the Ten Commandments. As human civilization has evolved, ideas of morality and rationality have evolved also. The notion that pork is unclean taught by some faiths is or was a way of preventing certain parasites and diseases. With our modern meat processing techniques, we have diminished much of the danger in eating what was once considered unclean. We don't need to hold onto archaic traditions as long as we stay true to the initial core values. There are sure to be certain rules and regulations that seem to be irrational or observed, but usually there is an underlying reason for their existence. As we progress through human existence, we must take the moral high road if only for necessity of co-existence. We understand the core values of morality and we thrive to use reason and understanding when we deal with life's issues. The principle reason we do this and try to live our lives in a proper fashion is to hopefully ensure our place in a positive second existence. We have examined our thoughts, beliefs and traditions of an afterlife from the beginning of human existence. We have pondered the existence of God and our purpose in the universe. We know what it's like to live. We need to know what it's like to die.

In our present society we are uncomfortable discussing death. The subject of death and dying is considered taboo. We tell children that "Grandma is just sleeping." We hear about death and dying every night on the evening news, but we don't stop to consider the complexities of dying. We buy life insurance to cover expenses of our death, but we don't think about actually dying. From the moment of our birth, we are in a race toward our final demise. We all will die. It is unavoidable. There is no religious ritual or medical procedure that will prevent it. We must understand that death is a biological process in the same way being born is a biological process. Dying is also a psychological process, a function of the brain. Only humans are afraid to die. Ernest Becker wrote in *Denial of Death* that it is "the basic fear that influences all others, a fear from which no one is immune …"

Some people so fear death that they will do anything to avoid it. They spend fortunes in an effort to stave off the inevitable. They considering having themselves frozen in the hope that someday they'll be revived and live forever. Death follows life and no one has ever avoided it. As we grow older and start to reflect on the aspects of death, we have a tendency to imagine the worst. In dealing with terminally ill patients who have been notified about their conditions, there are five stages of coping with their pending demise. They are denial (shock); anger (emotion); bargaining; depression; and acceptance (increased self-reliance). These stages were classified by Elisabeth Kubler-Ross after interviewing more than 200 terminal patients. She is a psychiatrist who has dealt with the reactions to death and dying by terminal patients. Not all patients go through all of the stages or go through them in any particular order, and some achieve one stage and are there until they die, but the five listed are the most common.

- Denial: The shock of being told they are going to die can create in a person feelings of loneliness and guilt. This can also lead to internal conflicts and thoughts of meaninglessness. This is usually temporary and they soon move into another stage.
- Anger: Why me? What have I done to deserve this? These usually are reactions to being told they are terminally ill. The best thing to do is to be sympathetic and try to reason through the experience.
- Bargaining: Eventually they realize what is going on. They try to determine if there is anything to delay the final ending. "If I pray hard, can I last until Christmas?" "Please let me live so I can see my new grandson." This stage usually doesn't last long and most promises go unkept.
- Depression: After surgeries or treatments, the patient soon realizes that they are weaker and unable to function as they would like. They become depressed because they feel they are helpless and unable to take care of daily affairs. They are becoming a burden.
- Acceptance: This is almost always the last stage in the progression. With help from professionals and family members, they are no longer angry and less depressed. They understand what is to happen and accept it, though with some reservations.

It is important to understand the psychological aspects involved with the dying process. We have never been taught how to die. Learning to die should be associated with learning how to live. If we can better understand the mental and biological processes, then we can reduce the fears involved. As humans, we have great difficulty dealing with the concept of the end of our existence. We have never known non-existence. We have no point of reference other than what we know as reality. When we do contemplate death, we tend to rationalize and try to find some purpose in death and dying. Regardless of any beliefs when we die, we are no longer part of this reality or life as we know it.

As we prepare to die, we realize that the only way we can still be part of this reality is in the memory of others. We can leave mementos and items that show we actually existed. Ultimately, our possessions tend to lose meaning and the relatives and loved ones who cherished our memories pass on themselves, and unless we imprint ourselves upon history we become forgotten and cease to exist. While we are alive, we are part of reality. Sounds rather bleak and disturbing, doesn't it? While we feel remorse about leaving this reality and leaving everything we know or have known behind, we start our new existence in an afterlife. There are stories told by survivors of near-death experiences that they see or feel the presence of relatives or loved ones that have passed who are there to guide them to the realm of the next existence. Even though we can cease to exist in the present reality, we will continue to exist in the next.

We have taken our journey through human history. We have seen how intelligent we have been since the beginning. We have examined our beliefs in an afterlife in cultures throughout time and all over the world. We have pondered the existence of god and considered the aspects of dying. If we are to understand what happens when we die, we must have an understanding of the physical laws of the universe. Any concept of an afterlife has to adhere to such laws. We next delve into the fascinating world of physics.

CHAPTER 6
What's Physics Got To Do With It?

Having now been ghost hunting in various locations for several years, I believe I've captured some interesting evidence to support the existence of the paranormal. I have an EVP and numerous photographs that show orbs, shadow people, mist forms, full-bodied apparitions and what could be called ghostly ectoplasm. I've heard disembodied voices. I have the evidence to show that there is life after death. With this knowledge I decided I needed to try to understand the mechanics behind this. There might be a way to understand what happens when we die. Having explored cultural and religious beliefs about the afterlife, I wanted to know how physics might shed some light what happens after our physical bodies die, if that was at all possible.

Any beginning to our understanding of physical laws should start with an examination of Albert Einstein and his general theory of relativity. Einstein developed his theory from a fairly simple experiment. Consider that you are in a small room without windows. If the room is moving or accelerating, you can't tell if it actually is or how fast it is moving. The principles of special relativity state this. The idea of any motion or acceleration must be made by external observation. If you are in the same room and there is a force pressing you to the floor, you can feel it. This occurs whether or not you have any external observation. We call these forces "G forces," which are the effects of Earth's gravity. At the Earth's surface there is a gravitational force of 1G. Let's put a scale in the room. When you stand on it, it shows your normal weight. Is there any way of telling if you are on the Earth's surface or accelerating at a rate of 1G? Einstein considered this and determined that the answer must be "no." There must be a relationship between acceleration and gravity. A localized gravitational field is indistinguishable from accelerated motion. This is known as the "equivalence principle."

If there is no difference between an accelerated observer in the absence of a gravitational field and a non-accelerated observer with a gravitational field, then we can always assume that the latter perspective is valid. We can say that all observers, no matter their state of motion, can consider themselves at rest as long as they include a suitable gravitational field to describe their immediate environment. Thus Einstein generalized his ideas of relativity.

We can use the equivalence principle to make another observation. If we are in the same room aboard a spaceship and we are standing against the walls and we point a laser beam at the opposing wall, we notice something odd occurs. In the amount of time it takes the laser to reach the other wall, the spaceship would have accelerated under the light beam. What we see is that the beam hits the wall slightly lower than it should. The motion of acceleration makes the beam curve. The beam will follow the course that takes the least amount of time between the two points. We would assume that the shortest time between two points would be a straight line. In actuality, if a beam of light is bent by a gravitational field then the shortest time between any given points would be a curved line. We can conclude that if we have a gravitational field then space is curved.

We can use the equivalence principle in another experiment involving time. Consider that our spaceship is somewhere in space and is using its motors to keep a constant position relative to the earth. If we turn off the motors we are immediately stationary in contrast to the Earth. We then start to accelerate due to the influence of Earth's gravity. The ship is in free fall and we become weightless. If we take our laser and attach it to the ceiling pointed downward and we fire it at the exact instant the motors were turned off, we would see that if we were directly beside the laser we could measure a certain frequency of the light. This would be the "standard" frequency of the laser. If someone on the floor measured the light it would have the same frequency. The conditions in the ship are the same because there are no fields present.

Now let's say we have someone on the Earth's surface as an observer. They would get the same "standard" frequency as the laser left the ceiling, but by the time the laser beam hit the floor the ship would have accelerated in the Earth's gravitational field, and is moving downward at a fair amount of speed. What the observer who is stationary relative to Earth's gravitational field will measure is a higher frequency. This is the same effect observed when an ambulance goes by and the siren seems to power pitch as it passes. This phenomenon is known as the Doppler shift. The energy being given off as something moves toward an observer is squeezed and its frequency seems to increase. This is referred to as being "blue-shifted." The same energy seems to stretch as the item moves away; the frequency appears to decrease and therefore is referred to as being "red-shifted." In simplified terms, if light shines downwards in a gravitational field it has a higher frequency (blue-shifted). If light shines upward in a gravitational field it has a lower frequency (red-shifted). This is a result of the equivalence principle.

Another surprising observation was made when we recognized that atomic oscillations that emit laser light are extremely accurate clocks. A clock can be defined as anything that repeats an action in a well-defined constant rate. It has been observed that time passes at different rates at different places in a gravitational field. It has been measured that time moves more slowly in a strong gravitational field than it does in a weak gravitational field. Take, for example, the U.S. atomic standard clock, which is accurate to one microsecond per year and is located at Boulder, Colorado. Its counterpart is kept at the Royal

Greenwich Observatory in England. The U.S. clock is located about 5,400 feet above sea level, while the Greenwich clock is roughly about sea level. The Boulder clock, because of its higher altitude, is in a slightly weaker gravity field and gains five microseconds per year over the Greenwich clock. GPS units must be calculated to account for this. This shows that a gravitational field not only warps space, but also warps time.

Einstein showed that space and time are interconnected and stated that space and time are variations of the same thing. We can conclude that a gravitational field can warp space-time. Einstein also argued that gravity is actually a warp or distortion in the geometry of space-time. The observations made by Einstein caused us to look at things differently in the cosmos. He showed that the Earth does not orbit the sun in accordance to some force, but actually follows the straightest path it can in space-time because space-time is curved due to the mass of the sun. Imagine placing a bowling ball on a bed mattress. The weight of the ball creates a sag in the fabric of the mattress the same way that the sun creates a sag or warp in the fabric of space-time. The Earth and planets follow the cusp of the warp and are balanced by inertia. The idea that space-time can be curved is a direct result of the equivalence principle.

In his theory of general relativity, Einstein provides us with a detailed description of gravity. The equations of general relativity provide us with a precise way to calculate the curvature of space-time created by a known distribution of mass and energy. With this we can calculate the trajectory of a particle through a curve in space-time. The theory also has the ability to predict the existence of black holes and gravitational waves, and accounts for the orbits of heavenly bodies. It gives us the best understanding for studying the universe. Einstein also introduced the idea of general coordinate invariance in which the theory looks exactly the same no matter what the coordinates of space and time are. When we combine ideas of general coordinate invariance and the equivalence principle, we have the essence of Einstein's mathematical concepts of geometry.

In the area of physics there are basically two theories – Einstein's general theory of relativity addresses the behavior of space and time, then we have quantum theory that addresses the interactions of subatomic particles. There are strange occurrences in the world of quantum physics which will come into play later when we examine ghostly encounters. As we proceed, we are gathering pieces of a puzzle that we hope to assemble into a picture that explains the biggest question involving humanity. There are many pieces from many places that seem unrelated to each other or the subject at hand, but as in art many colors make a painting.

There are basic properties of quantum physics. Matter consists of particles and the properties of particles are usually limited to a number of allowed values. The state of a quantum object is completely well-defined. The theory predicts how the state changes. The unpredictability of the theory comes into the equation when we try to rotate this to observations.

A single quantum state can correspond to multiple positions. Only so much information can be obtained from a quantum particle. If you know exactly where a particle is and you have no information regarding how fast it's going, then when you determine the speed you lose track of its position. If two particles are exactly alike you can't be sure if they have changed places. It is the sameness that governs the particles' behavior. If you view two conjoined particles separately they do not seem interconnected, but if you look at them together they form a pattern. The pattern has a complexity that is unexpected. They seem to be coordinating their pattern across the space that separates them. If you are trying to observe a quantum particle and you can't tell exactly where it is, you must assume that it is everywhere it can be at the same time. If you have many particles they will end up in various locations at random. They will go anywhere and do anything that the theory allows.

A very interesting experiment was developed to measure photon interference. Light photons were sent through a slit and collected on a screen located past the slit. The interference would be when photons interfere with each other and would hit the back plate at different angles, like paint from a spray can. When they used two slits the photons should dilute because they are being spread out by the slits. This didn't happen. If they fired only a single photon at the two slits, they saw that interference still occurred. This seemed impossible because a photon can't interfere with itself. It seemed as though the photon was going through both slits at the same time. Not only was the photon in two places at the same time – the photon was also in two different times at the same time. Each passing through a slit constitutes its own movement in time. Weird stuff, huh? A particle takes all paths available to it.

One process we can look at has the same effect. If we look at a wave traveling down a road and it reaches a fork, the wave splits and when the road comes together again the wave does the same thing. The fact that the wave splits lets it continue on its journey. The photon acts like a wave because it takes both paths to get to its intended target instead of one or the other.

Another interesting and what some have referred to as a "spooky" aspect of quantum mechanics is what is known as entanglement. It seems as though scientists like to refer to scientific properties in single-word descriptions. When two systems – usually particles – enter into a temporary physical interaction due to known forces between them and then separate, it is by the interaction of the two states that they have become entangled. If we take the pair of entangled particles and conduct measurements on one of them, say we measured the spin of one particle, this can affect the second particle no matter how far away. No wonder they call it spooky.

Entanglement can even exist between two separate properties of the same particle, which might include such properties as spin and momentum. What we have are single particles or pairs that can be entangled due to any combination of their quantum properties. The strength of the quantum link can range from partial to complete. When a change is effected on an entangled particle it can occur at any distance and is instantaneous. Even if the particles are at opposite ends of the universe, the change occurs immediately, which shows that the effect is not governed by the speed of light.

On the other hand, the degree of entanglement between spin and momentum in a single particle can be changed by increasing its speed, which is referred to as "boosting." This boost can actually create spin or momentum entanglement — or both — between two particles that neither one had to begin with. The boost in speed also can enhance the entanglement related to spin, but this decreases momentum, or it can decrease both spin and momentum.

There are various strange effects involved with entanglement. The spin of one particle can be entangled with the momentum of the other without having any entanglement between the two spins or the momentum of either. A pair of particles can have a grand entanglement that includes all the possible combinations of spin and momentum that stay constant when boosted. There also can be a change in the reference frame that alters the various components of the overall entanglement in such a way that is not understood very well. Pre-existing overall entanglement can be concentrated in the spin, which seems to be creating spin entanglement from nothing. This effect may offer a simple way to manufacture entangled particles for teleportation and ultra-fast quantum computer experiments.

There are areas of study being developed to combine quantum entanglements and gravity, which is an essential part of the theory of relativity. A strong gravitational field can actually create entangled particles and those that break free are called Hawking radiation. Gerard Milburn of the University of Queensland Australia said, "It would be very nice if this could be turned around and Hawking radiation derived as a consequence of quantum information in curved space-time." Spooky!

Okay, we have looked at the effects entanglement have on particles. We need to know what particles are. There are basically two families of particles. The first is the electron. The electron is a particle of matter. The second is the photon. The photon is a particle of force. Electrons also are known as fermions. There are several members of the fermion family, which are divided into generations. The progressive generations unite to form heavier particles. The families include quarks and leptons. Each type of fermion has an opposite or anti-matter equivalent. The anti-matter particle has the same mass as the standard particle, but has the opposite electrical charge. The electron's anti-particle is known as the positron. If the two meet, they annihilate each other in a burst of energy. The known universe is made predominantly of matter.

Photon particles also are known as bosons. Photons are the particles responsible for all electrical and magnetic phenomena. The photon is the unit of energy within a light wave. It's kind of like a wave itself, and by combining them, we can get a light wave that we can see with the naked eye. Photons also create electromagnetism. This is the prominent force involved in everyday life. It is the engine behind electricity and magnetism. They also create chemical reactions, generate light and keep structure from collapsing under their own weight.

To generate a force, a particle has to be electrically charged. Particles that have an opposite charge attract each other. Ones that have identical charges repel each other.

Another aspect is the weak nuclear force. This force is more obscure. It has a very short range, which is confined to the nucleus of an atom. It is the force that drives the unification of fermions. Then we have the strong nuclear force. This is what holds together the nuclei of atoms. The protons inside the nuclei of atoms are all positively charged, which means they would repel each other if something did not hold them together. The strong force, like the weak force, has a very limited range.

One thing about particles is that every one of a given type

behaves in the same way. They could have different amounts of energy, but their properties are the same. This is contained in the standard model. According to the model, particles exist in a field that fills space somewhat like a force field in science fiction. The fields stretch across space and exert forces. The best-known example would be the magnetic field we can't see but that is all around us. Another example would be an electric field that creates electric sparks. Actually, the electric and magnetic fields are aspects of the same thing – electromagnetic field. We think that fields come from an object such as a magnet, but the standard model suggests the opposite. Fields are said to be fundamental and particles come from them. Whenever a field gains energy it begins to emit particles. Each type of particle has its own individual field, and the photon is the particle unit of the electromagnetic field. Photons are created or destroyed whenever energy is added or removed from the field. The electron is the particle unit of the electron field. Every electron is the same because they are all generated by the same field, which stretches throughout the universe. This field is permanent even though individual electrons come and go. Even in the vacuum of space the field is there. It never completely goes away.

So you now have an understanding of some of the key elements of physics, so as we continue you can refer back and say that this actually might make some sense. We still have a ways to go, but it gets a bit interesting, as we shall see. We have looked at Einstein's theories and have tip-toed in quantum mechanics. Now it's time to examine the line of inquiry that binds the two — string theory.

CHAPTER 7

A Symphony of Strings

String theory proposes that matter, space, force and time consist of vibrating strings. This is thought to be the most probable candidate for a unified theory of physics. Atoms are made of subatomic particles called protons, neutrons and electrons, and these are made up of even smaller particles called quarks, and even smaller than that are strings. If an atom were the size of the solar system, then the string would be the size of a tree. The length, curves and vibration of the string determines the nature of the atom.

Imagine that the strings are tiny guitar strings and each type of particle corresponds to the string playing a certain note. What we would have would be a symphony of unimaginable intricacy. String theory not only unites types of particles, but also the way they behave. Actually, string theory isn't a die-cast theory such as the theory of relativity where specifics are verified, but more of an approach to the explanation of processes. It is more a set of ideas than a full-fledged theory[4].

---

[4] It's probably worth noting that unlike the popular use of the word theory to mean

Physicists have found different ways to observe strings that simulate others under certain conditions. String theory combines the special theory of relativity and quantum principles, but uses strings instead of particles. In 1968, Italian physicist Gabriele Veneziano was studying an equation and observed that this could possibly explain particle reactions that involved the strong nuclear force. Several of his colleagues later determined that the equation would make sense if the particles were connected by strings. An idea was born. This idea was eventually expanded to include not only the strong force but all forces, including gravity.

For several years, the idea was put on the backburner. It was the stepchild of physics that no one wanted. Few physicists pursued it, and those who did had a hard time trying to be recognized as legitimate scientists. In 1984, there were a series of breakthroughs that have become known as the First String Revolution. Soon after, Edward Witten, a noted particle theorist from Princeton University, got involved and in the matter of a few weeks strings came into their own right, but after a few years, the revolution ran out of steam. The theory branched off into seemingly incompatible versions. It was thought that with inconsistent variations of the same theory, the end was near. The hopes for a unifying theory were evaporating. Then in 1995,

---

speculation without proof, the use of the word theory in science is more specific and is used to describe an idea that has been tested and retested and is widely accepted as true based on the evidence.

Edward Witten gave a lecture at the University of Southern California and set forth the principle that the different versions of string theory actually were compatible. This is considered the Second String Revolution.

So, we know that strings vibrate and that they are incredibly small. There are other aspects that show almost magical properties. Strings can be open-ended or loop around themselves like a rubber band. They never sit still but are always vibrating. They can wiggle and wrap around things, twisting and turning. A string can vibrate one way and start vibrating in a totally different way. When this occurs, a particle can change from one type to another. It can even go from a particle of matter to a particle of force. More spooky stuff.

Two strings can exert a force on one another by using a third string, which is created by the strings dividing like a cell. The strings determine the strength of the force. One way this is done is to pulse somewhat like a heartbeat. The rate or strength of the pulse determines the harmony of the strings, which affect the strength of gravity, electromagnetism and nuclear forces. The strength of these forces are the dynamics of strings.

One thing in particular about string behavior stands out and draws the attention of many physicists. A closed string can vibrate by expanding in one direction and contracting in the other direction, then contracting and expanding in the opposite direction, back and forth continuously. This type of vibration or pulsation mimics what gravitational waves do. It can be assumed that when a string is acting in such a way it is behaving as a graviton. Sending off gravitons is how particles exert gravity on themselves. When we add gravity into the mix, it becomes harder to maintain particle consistency. The result is that the string vibrates in a curved space-time. We must add an additional constraint to maintain order. This, it turns out, is the general theory of relativity. String theory cannot exist without general relativity. The result is a unification of theories.

Using strings we have made an attempt to explain particles and the forces of nature. Primarily, the objects in question have been one-dimensional. Physicists have found that the theory also predicts multidimensional objects. Among these objects are two-dimensional membranes that stretch across space and objects known as branes that include lower and higher dimensional versions. As a point of reference, a 0-brain is a point, a 1-brane is a type of string, a 2-brane is a sheet, and a 3-brane is a solid body. The higher the dimension, the heavier they are, which takes more energy to create, which makes them rare. Branes exert forces on each other and can orbit one another. One type of brane called the D-brane can attach to the loose ends of strings and flap like a sheet in the wind, or the brane can hold down the string and limit its ability to vibrate. The 0-branes can fly around and actually cause a string to vibrate. D-branes are sheets that open-ended strings can attach to. Hopefully we have gained an understanding of strings and string theory. All of this brings us to bigger and better things.

At the beginning of our known universe, there was a massive, intense explosion referred to as the Big Bang. In the first trillionth of a trillionth of a second, a mysterious anti-gravitational force caused the universe to expand at a speed greater than the speed of light. This can happen because space itself is expanding, which does not conflict with Einstein's theory of relativity. This expansion is known as inflation. This expansion is still going on today with stars and galaxies along for the ride. Imagine raisins embedded in bread dough. As the bread rises or expands, the raisins expand with it, moving farther apart from each other. This is what is happening to cosmic features. It was thought that galaxies were moving away from each other due to the explosion effect, but they actually are moving due to the expansion of space.

No one knows exactly what caused the inflationary period to start. It could quite possibly be happening in other parts of an infinite universe. Big Bangs might be common across infinity. If this is so, then possibly a small area of a universe could suddenly inflate and sprout another and then another. We would then move from a universe to a multiverse. This theory is known as chaotic inflation. What this does is create the idea that this process is eternal and that the universe or multiverse is infinite with no beginning and no end. This also suggests that our known universe may at one time have been a sprout or spawn of another universe.

These universes could very well have many different physical laws and may contain an infinite variety of life forms.

Another interesting idea associated with this is that if the universe has no beginning, then time would be timeless. One aspect of relativity theory states that space tends to grow and pinch off to create new areas. It would be an anomaly if other universes didn't exist. According to string theory, the space within a universe has an irregular shape when the universe comes into being. The energy emitted by space and the influence of the shape of this space determines the type and existence of particles and the physical laws that govern them. The number and effects of dimensions also are determined in this manner. This creates the possibility of infinite realities.

Another way to look at this is if a reality can exist then it will exist. In string theory and in the standard quantum theory, all possible shapes of interdimensional space behave as though they actually occur. If they behave as though they are real then we must conclude that they are real. Theory concludes that they could be levels of multiverses, the first containing universes that are beyond our range of vision where the density of matter varies. The second would contain bubbles in an empty space, the types and properties of particles and forces vary. A third possibility contains quantum possibilities of the types and properties of particles and forces, and the last contains everything that could possibly exist.

Theorists are at odds as to the specific fundamental laws that could govern multiverses. Some argue that they must be mathematically consistent. Others say that any possibility could exist. One thing that is present in our universe, and I would think would have to be present in any variation of multiverse, is some form of cohesion. If our universe had formed differently, we could have ended up in a vast array of subatomic particles. A slight variation in the strength of electromagnetic energy or in the weak or strong nuclear forces would have changed our existence greatly. As with any form of multiverse, whatever laws govern that existence or whatever particle form or force level exerted must have some kind of cohesion. It has to form to the natural state of its environment. It is the natural state of our environment to be what we are and for our universe to exist as it does.

The standard model contains particle masses that seem as though they were chosen at random, but the common theme is that they are connected to the universe. There are an infinite number of multiverses. We live in this one because it is conducive to our existence. If we lived on the only planet in our universe, then the habitability of this planet would be an undeniable fact or a consequence of the laws of physics. We know this isn't true. There probably are millions or billions of planets in our universe. We live here because life on this planet has evolved in accordance with its environment. So if a particular universe forms as to the nature of its environment it must also form as to the nature of all multiverses. This is to say that it can have its own unique properties but it has to be in balance with other universes. There has to be an overall balance to maintain order.

We can take into account the influence of dark energy. This is an unknown form of energy that is causing the expansion of the universe to accelerate. Dark energy doesn't seem to have been around long. It is believed that it began to exert its influence somewhere around the time our solar system was formed. If it had been around earlier it could have kept galaxies from forming or stars from igniting. The density of dark energy must be incredibly small. According to string theory, dark energy comes from the same process that created inflation or the expansion of space. This process is believed to come from the actual shape of extra-dimensional space. The second level of parallel universes contains varying amounts of dark energy. The vast number of those universes suggests an adequate amount of dark energy. The more universes there are, the greater the possibility that some will have the proper amount of dark energy produced at the appropriate time for galaxies and stars to form, and life would arise at various locations.

Not all properties need to be balanced to maintain a universal order. Take for example the asymmetry of matter and anti-matter. Matter doesn't have to be so far more abundant anti-matter for us to exist. There could be millions of times more anti-matter than exists here and life could still exist. It could be that it's the nature of this universe for anti-matter to be so rare. It is said that order naturally decays into disorder where disorder stays that way. For it to continue, the universe has not reached disorder yet but it seems that the universe is more orderly than it needs to be for our existence.

CHAPTER 8

Into the Wormhole

Okay, we have parallel universes with stars, planets and physical laws unique to the environment of their formation. We can now go through the looking glass as Alice did and visit these wonders. Albert Einstein first thought that black holes were only possible in theory. He could demonstrate with mathematics that there could be what is called a wormhole at the center of one. According to general relativity, wormholes can be formed when a black hole bends or sags the fabric of space-time enough to produce a hole or tunnel to another point in space-time. So what's on the other side? Einstein theorized that there was a mirror universe at the end of the wormhole or tunnel. What we would have would be a parallel universe resembling our own. At the end of the wormhole you would exit at a white hole.

A wormhole is basically a shortcut through space-time. If you lived in an apartment and wanted to visit your neighbor, you would have to exit yours, go down the hall, and into your friend's apartment. If you had a doorway to your neighbors you simply would walk through the door. This is the same with a wormhole.

Instead of walking, say, 30 feet, you only have to go 3 feet to get to the same place.

Einstein has shown that space-time is curved. Instead of following the curve to reach a destination, a wormhole could bypass the curve and provide a path directly to the destination. When subatomic particles were first observed they tended to disappear and reappear in a mysterious way. It is quite possible that they were entering and leaving a parallel universe through subatomic wormholes. Theorists speculate that there could be what they call quantum foam, which consists of particles and subatomic wormholes. Particles could move between parallel universes and through time.

When we look at Einstein's theory of relativity it states that space and time are variations of the same thing. Any wormhole that connects two points in space should connect two points in time. This opens the possibility of time travel. The ideas of time have changed and evolved over the centuries. To Newton, the idea of time was that it followed a straight and unchanging course from the past to the future. Einstein introduced the idea of warped space. If space was warped, then time must be warped also. Not only was time warped with space, but flowed somewhat like a river that sped up and slowed down. This river could possibly bend back on itself and have currents, whirlpools and even forks in different directions. These forks could branch off to form different parallel universes. One branch would continue and create the history of time that we have and another branch could form an alternative existence such as one where the Germans won the Second World War or where Lincoln survived the assassination attempt. It is thought that all possibilities exist, but we can only comprehend the existence that is observed. If you remember a science fiction TV show from the 1990s called *Sliders,* where a group of individuals used a wormhole to travel to parallel Earths with different historical outcomes, this would be the concept described.

In a recent publication by Stephen Hawking, as summarized by Life Science senior writer Clara Moskowitz in a 2010 article, "One consequence of the theory of quantum mechanics is that events in the past that were not directly observed did not happen in a definite way. Instead they happened in all possible ways. This is related to the probabilistic nature of matter and energy revealed by quantum mechanics: unless forced to choose a particular state by direct interference from an outside observation, things will hover in a state of uncertainty. For example, if all we know is that a particle traveled from point A to point B, then it is not true that the particle took a definite path and we just don't know what it is. Rather, that particle simultaneously took every possible path connecting the two points. Yeah, we're still trying to wrap our brains around this. The authors sum up: 'No matter how thorough our observation of the present, the unobserved past, like the future, is indefinite and exists only as a spectrum of possibilities.'"

In an attempt to unify gravity with quantum mechanics, the idea of imaginary time comes into play. Imaginary time is indistinguishable from directions in space. If one goes in one direction, then they can go in another direction; correspondingly, if one goes forward then one can go backward. There should be no significant difference between going forward or backward or choice of direction. In dealing with the properties of particles the laws of science do not differentiate between past and future. The laws that govern matter remain unchanged under all normal conditions.

As we look at time itself, we can determine that there are what are called arrows of time. One arrow is the thermodynamics arrow of time. This is the direction of time in which disorder would increase. Then we have a psychological arrow of time. This is where we actually are aware of the passing of time. Another is the cosmological arrow of time. This takes into account the expansion or inflation of the universe.

In the thermodynamic arrow, the universe would have begun with a rapid rate of expansion. During this expansion the variations in density would have been small but would start to increase. Areas where the density was a bit higher than others would have their expansions slowed by the gravity of added mass. These areas would quit expanding and collapse, forming stars and galaxies. The universe would have started in an orderly state and then become disordered due to the variation of cosmological processes. This would explain the existence of the thermodynamic arrow of time.

The cosmological and psychological arrows are interconnected because as the universe is expanding, that's where we get our awareness of the passing of time. Time is expanding with our universe.

When we look at the laws of science, they do not differentiate between the forward or backward directions of time. However, when we look at the arrows of time they do. All of these follow the same direction of going from the past and into the future. The observable direction of time is one way. Even though the cosmological and psychological arrows are associated with the expansion of our universe, time would not reverse if the universe started to collapse. The concepts of the arrows of time are contained in the elements of real time. Real time can be said to start at the Big Bang or the beginning of our universe. Real time begins the second our reality commences. The concepts of real time are that there is a beginning and an end. Real time ends with the collapse of our universe and therefore the collapse of time.

To comprehend time before the Big Bang we need to refer to imaginary time. As we have seen, imaginary time would have no boundaries. With imaginary time there would be no beginning and no end of time. The idea of imaginary time could very well be the actual nature of time. The notion of real time is an expression of what we observe as the passing of time as we know it. The idea of real time exists only in our minds. If one asks the question which is true — either imaginary or real time — the answer would be whichever would offer the best description to the question asked.

When we say that imaginary time would have no boundaries we should actually say that space-time would have no boundaries. Imaginary time would exist in an infinite space-time. Using these, we can determine the likely properties of our universe. We can calculate that the universe is expanding at about the same rate in all directions using the amount of mass contained within. It seems that using the no boundaries possibility, the equation seems to be true. This is also observed in the background radiation that shows it has about the same intensity in all directions. At the beginning, our universe would have gone through a period of rapid expansion. Any variations in uniformity would be exaggerated. Gravity would cause more dense areas to slow their expansion and start contracting. This would lead to the formation of stars and galaxies. All of these could exist because there are no boundaries to restrict them.

It is very possible that there is no beginning or end to the universe. Space as we know it is contained in our universe and is expanding with stars and galaxies going along for the ride. Time itself would be infinite as well as existence beyond our bubble of space.

CHAPTER 9

Time in a Bottle

When it comes to our perception of time, it is interesting how we perceive that time is biological. The part of our brain that could be termed our "biological clock" is the basal ganglia[5] located deep within the base of the brain. This and the parietal lobe[6] located on the surface of the right side of the brain are the areas critical for our ability to keep time. These areas have nerve cells that contain the neurotransmitter called dopamine[7]. In diseases such as Parkinson's disease, there is an abnormal reduction of dopamine in the basal ganglia that creates problems with time perception. There is an improvement in this area when dopamine levels are increased.

Biological clocks exist in virtually every living thing because they are essential for survival. Plants need to spread their

---

[5] A cluster of neurons that function as a single unit and are associated with functions such as voluntary motor control, learning routine tasks, and cognitive and emotional functions. http://thesciencedictionary.org

[6] One of the four major parts of the cerebral cortex, which is the main body of the brain. The parietal lobe receives and processes information from the senses and is associated with the ability to form words and thoughts. http://science.education.nih.gov

[7] A natural chemical in the brain that helps control the reward and pleasure centers and regulate movement and emotional responses.
http://www.psychologytoday.com/basics/dopamine

leaves to get the optimum sunlight for photosynthesis; the eyes of fish need about 20 minutes to adapt to light or darkness they know when to start the process as not to miss a feeding opportunity or to become food themselves. Something that I think is really cool is that you can actually plant a flower clock. You can take different flowers and plants that bloom or open at specific times and arrange them in such a way that they bloom in accordance with the hours on a clock and could show the exact time.

Certain people can be called morning people and some are considered evening people. This could be hereditary. If so, then my oldest daughter didn't get this gene. Having worked morning and daytime shifts most of my working life and being an avid hunter has made me a definite morning person, but my daughter doesn't seem to lean that way and is content with the evening lifestyle. To each their own.

The human biological clock doesn't seem to be set to the length of the day as per a 24-hour cycle. It has been noted that we seem to exist in a cycle closer to 25 hours. People spending extended time in caves and people living in the far north — where there are seasons of constant daylight and constant darkness — would fall asleep and wake at regular times, but their sleep cycle would gradually shift later every day.

Before the Industrial Revolution, most people followed the natural rhythm of day and night. They didn't have much artificial light, and relied on candles and oil lamps that were not very efficient as light sources. During this era humans tended to sleep between nine and ten hours each night.

It was common during the Middle Ages for people to have segmented sleep. People would go to sleep at sundown and wake up for an hour or two in the middle of the night and return to sleep until dawn. This was known as sleep one and sleep two. During the awake time between sleeps, people would visit neighbors or do some cleaning or take the opportunity to have sex. There is a theory that this is where our fear of darkness comes from because during these midnight visits highwaymen would rob and victimize people venturing out into the night. These fears along with superstitions about witches and goblins led to a deep-seated, yet unfounded, fear of the darkness.

With the advancement of the Industrial Revolution and the invention of the light bulb, we as a culture had to learn to reset our biological clocks so that we slept through the night because we were working regular jobs in manufacturing and related endeavors, and we had the need to sleep through the night and be awake all day.

Our perception of time implemented by our biological clocks is vital to our survival. We know when to sleep, when to eat, when to perform the functions necessary for us to live. That is what our biological clock does. In organisms that have much shorter lifespans, the perception of time has to be much different.

During our perceived lifespan we go through life at a pace to where the passing of time is consistent with the time we can exist. In an organism that's lifespan is maybe only a day or two, then the time perceived is in proportion to the lifespan. To this organism, it has lived its full life and that day or two may feel like 80 or 90 years.

The perception of time is also known as a sense of time. This can be considered a sense along the same lines as a sense of sight or hearing or touch. With our sense of sight, we have evolved so that we see what is necessary for our survival. When we see, we only see certain spectrums of light. We don't see many wavelengths of ultraviolet because it is not necessary. When we hear, we only pick up a certain range of decibels, unlike a dog, which picks up a wider range and can hear sounds above what we can, hence the dog whistle that is audible to a canine but inaudible to a human. We don't have the sharp, long-distance vision of an eagle, the hearing of a dog, or the olfactory abilities of a deer or bear. What we do have is adequate for our survival.

It is the same with our sense of time. Our biological clocks give us a sense of time adequate for our survival: when to eat, when to sleep, etc. As with our other senses, which pick up a mid-range of the overall spectrum of available information, I would suspect that our sense of time is the same way. When we take into account that our sense of time is basically the ticking of the clock and the passing of days, weeks, months and years — while other organisms perceive time differently — then there must be aspects of time of which we are not aware. We have developed instruments to show us different spectrums of light and sound, but not of time. The only information we have about time is what we can observe. Einstein proposed theories about time and space, and we can experiment with speed and time variations, but to observe time beyond our own initial perception is difficult. When we become Homo quantumus after we leave our organic bodies, we don't have the same biological time clock to govern the way we perceive time. We can observe time in its true form. When we exist in time-space instead of space-time, the full spectrum of time will be at our disposal. Existing in time will be our natural state.

CHAPTER 10
Pulling It All Together

As we have seen earlier, virtually every culture all over the world has had a belief in an afterlife. We have shown that Homo Heidelbergensis had a form of ritual burial at least 350,000 years ago. It seems as though we have the ability to tap into a universal knowledge that lets us be aware of an afterlife without specific knowledge of it. I would think this is the same process that allows a flower to use bees to pollinate and continue their species without seeing them. They are aware of the bees' existence. We have a primal awareness of an afterlife, which carries with it a primal fear of death.

As we have seen with terminally ill patients, we have developed processes to rationalize and prepare for passing on. When the first cell divided two things occurred. First came the development of metabolisms and the beginning of organic life. The second thing that occurred is the transformation of energy from whatever source — be it solar or geothermal or whatever energy source was available — to life force energy. One of the laws of physics governing our universe states that energy cannot be created or destroyed, just transformed. When we transform from organic to energy beings we must enter into a parallel universe. We exist as organic beings in space-time because that is what our physical laws allow. It would be more physically possible to exist in time-space as beings made of energy or particulate. Since energy cannot beget energy, then there has to be an organic host to form and nurture the life force until it passes to the next existence. One would think that with the complexity of life and the evolutionary investment we should live to be at least a few centuries old.

I am reminded of an episode from the original *Star Trek* TV series from the 1960s. In the episode, the inhabitants of a dying planet went through a portal that allowed them to escape to the past, therefore saving the population. Instead of a portal, we have stories of going through or into the light. I would think that "into the light" would actually be into the wormhole. When scientists first observed sub-atomic particles they would appear and disappear at random. The thought was that they were entering and leaving a parallel universe through sub-atomic wormholes. If we are energy or particulate, it wouldn't matter if the wormholes were sub-atomic or not. There are also areas of quantum foam made up of particles and sub-atomic wormholes.

I don't believe our journey to a new universe once we become energy beings would be a one-way trip. We should be able to move from one universe to another using wormholes. We should be able to return from time to time if we want to check on loved ones or whatever our interest is. If we are attached to a person or thing, this could be the mechanism for hauntings to occur. Since we are from this universe some of ourselves would carry over to the next. Ghost investigators often will use the phrase that a ghost has "manifested." If so, then where did they manifest from? If the ghost wasn't there before then it had to come from somewhere. I would contend that they arrived in this time and space from what some call the other side by some means, probably a wormhole of some kind.

When we become Homo quantumus or an energy being, we become a self-contained energy field. We are an assembly of unified particles containing weak and strong forces. We are life-force energy. This could very well be its own form of energy contained within itself. The phenomenon of life-force energy has never been studied. We must have the ability to alter the density of particles. We seem to be able to shift from energy to matter. The difference between energy and matter is the density of particles. When the Big Bang occurred during the first microseconds, everything was energy and this condensed into various forms of matter. Einstein's equation of $E=mc2$ concerns the conversion of energy to matter.

I have taken pictures at reported haunted locations showing the various forms that ghosts or entities take. The forms that I have photographed are orbs, shadow or black mass forms, mist forms, recognizable full-bodied apparitions and what some call ectoplasm. If I were to guess what the nature of each form was I would say that an orb or ball of energy would be the form as it exited or traveled through a wormhole. The shadow or black mass form would be the assembling of particles to make a form. The mist form would be another form of the assembling of particles, possibly due to the form or availability of energy to be used. The apparition or most human-like form would be the most complete assembly. I would say that the persona of the assembly would show what the person had looked like or what it wants to look like as opposed to what the entity actually is.

I thought this might be a stupid question, but why do apparitions appear to wear? Clothes are not spiritual, they are inanimate. A clothed figure must be the way a spirit wants to be perceived. This would be especially true for soldiers or uniformed entities. There is a theory about ectoplasm that it is the residue left over by the manifestation of a ghost. This could just be particulate residue. With the ability to vary the density of energy or matter, a ghost or spirit would have the ability to move objects or to open and close doors that occur on some investigations.

Altering density also could give the entity the ability to form words from surrounding energies or particles such as air. I have an EVP that I didn't hear with my ears but picked up on my recorder. The voice seemed close to the microphone. There was an investigation conducted at the Lizzy Borden house in Fall River, Mass., where an EVP was recorded. There was a voice recorder on the bed and a camera some feet away. Both pieces of equipment picked up the voice with equal volume. It seems as though the voice recorded was omnipresent.

I also have heard a disembodied voice with my own ears. This could be attributed to sufficient energy and particles available to generate a hearable voice. The only thing we have to make theories or assumptions on is the information provided by first-hand observations of others, which usually are amateurs such as myself. We take this information and form our best guess scenarios. There has to be, at some time, valid established scientific study of these phenomena. The story of human existence will not be complete without this knowledge. The concept of life itself needs to be examined as the force for which it is. The life-force that animates us is one of the most powerful forces in the universe. Once it is established it will continue at any cost. We need to examine why there is so much what I call anti-life. Once a disease is eradicated another forms to take its place. Why would a disease kill its host, and thereby kill itself? A drive toward suicide doesn't sound like the most logical biological imperative, even for a microbe. I think this relates back to the law that energy cannot beget energy. Energy needs an organic host. Once the host is established to nurture the spirit or energy being, then processes need to be in place so the organic host will die and release the spirit energy being to enter into the parallel universe so that the cycle of life and metamorphosis can complete. It is quite possible that life forms ultimately were intended to be energy beings because energy beings could exist forever. As we have seen before, physics states that energy cannot be created or destroyed, only transformed.

Albert Einstein envisioned parallel universes as mirroring our own. In this synopsis, an infinite number of parallel universes would mirror each other in infinite ways. Every possibility would be represented. The universe in which we are intended to live as energy beings would be a time-space universe where we exist in time as opposed to space. As organic beings in this physical universe, we exist in space-time because that is what our physical laws allow.

Physicists say that time travel is possible but it would be very difficult to achieve because of our solid nature. When we look at the properties of time and space, we could possibly form an equation to compare one to another. If we said that a day could equal a mile and we took the time that the earth will exist, approximately nine billion years from beginning to end, then time would be over three trillion miles, which is about a quarter of the way to our nearest star — plenty of elbow room.

In this existence we have three dimensions of space. In a time-space universe there could be three dimensions of time. There could be time that goes front, back, up and down and from side to side. As this universe is expanding in space, the other universe would be expanding by the passing of time. As time progresses there would be more past in existence and the past would always exist. As we have stated before, time travel is possible in theory. Since this is true, then the past must always exist. If we would exist in time and the past always exists then we could exist forever. What a concept. That Star Trek episode where they escaped into the past seems prophetic.

We had discussed earlier the metaphysical properties of the existence of God. It would seem that this is the perfect solution to the question of how we could possibly exist forever. Our eternal existence also has the signature of a divine presence or intelligent design. In this universe we can go basically anywhere in space, within the limitations of our technology. In the other existence we could go anywhere in time. There would be no paradoxes in time-space existence. In a parallel universe that mirrors our own, any event in history would be happening at any particular time. In this reality you can go from New York to Los Angeles. If we exist in time, you could go from now to then. You might go from World War II to the Civil War and back again. If you can go anywhere in time then the future would probably exist. You could go from then to now, to tomorrow. Since the future doesn't exist to us then all possible futures could exist. The possibilities are endless. Like the ghost woman waiting for the train in the 1940s at the abandoned train station that the ghost investigators encountered, all we have is time.

CHAPTER 11

Why Hauntings?

I have had several family members pass away, and the previous owner of my house passed here — and yet I have never knowingly been visited by or heard from any of their spirits. I have to go to reportedly haunted locations to try to collect evidence of ghostly activity.

When I investigated my friend's house and recorded my EVP, the spirit was active that night but when I went back some months later I didn't catch anything. I don't think it was there that night. They must be able to enter and leave on occasion. When a ghost manifests it has to manifest from somewhere. I would think that if a spirit comes from a reality governed by time that they would not have the same linear sense of time we experience. We don't have a sense of space because that is our medium, so it would be safe to assume they are not conscious of time. The idea of ghosts using wormholes — even if they are subatomic — could have some validity. Some entities recognized as being demonic also could travel this way. They are probably from another parallel universe since there are an infinite amount of them. Demonic entities reportedly are beast entities that were never human. In my paranormal travels I actively try to avoid them.

We can look at investigations others have done to try to better understand paranormal phenomena. A ghost investigation team traveled to Texas to investigate a World War II-era aircraft carrier. The claims of activity were varied and all over the ship. On one occasion a blue ball of light was observed on the flight deck, first hovering flitting away through an open door. On another deck, a uniformed figure was seen leaving one door and entering another. In the same area, a disembodied voice was heard saying, "Hey." Another figure was seen walking through a wall where a door used to be. In an area below deck where a torpedo struck near the engine room, shadow figures were observed running around, and the sound of men screaming could be heard. In many areas of the ship, footsteps and voices could be heard and there was no one to be found.

During the investigation, two members of the group went into the power shop located on level three and reported encountered feelings of dread and an overwhelming urge to leave the area. As one investigator put it, it was like walking into a room where an argument had just taken place. After they left the area, the feelings subsided. There didn't seem to be an explanation for what had occurred. In a lower deck area, members of the group were investigating when they thought they heard footsteps ahead of them. The footsteps seemed to grow louder and louder as they proceeded down a hallway. Along with the sound of footsteps, they could hear an occasional banging noise. While in the hallway they glimpsed a shadow figure entering a doorway. When they reached the entrance where they saw the figure, no one was there. The figure was gone and so was the sound of footsteps.

As investigators were checking out an area on the first deck, they were startled by what sounded like someone dropping a tool or wrench. This sounded like it was very close to them. Right after they heard the metallic noise they could hear what they thought was a cough. As they entered the place where the torpedo had damaged the ship, they heard the sound of whispering in the distance and splashing noises as if someone was walking through water.

Another team went to the engine room where they conducted an EVP session. During this session they heard a woman's voice with their own ears but couldn't make out what it was saying. When they reviewed the evidence they had collected, they found their equipment had picked up what sounded like a man's voice saying something about the ship and then some breathing noises. They also caught a strange light that appeared briefly on a wall. From what they saw and encountered and the evidence captured by their equipment, they were able to confirm the stories of hauntings that were related to their experiences. All the evidence indicated this was what is referred to as a residual haunt.

According to paranormal investigators a residual haunt is basically the impression left by a traumatic event. Somehow a psychic or emotional impression is left, usually around limestone or quartz (none of these are on the ship) to be replayed at a given time by some unknown means. They say that there are no spirits involved, just visual memories of the event. I would tend to believe there are actual spirits involved rather than sense memories only. Although these materials could have properties like those described, the it just doesn't seem logical to me that we typically only see human spirits acting out historical events at these kinds of haunts. If traumatic events leave impressions, then why don't we see images of a mountain lion brutally devouring a deer or a saber-toothed tiger destroying prey from the distant past? We should see images of car crashes playing like movies in quartz or limestone-rich areas.

What actually is happening in a residual haunt is an area that needs legitimate scientific inquiry and answers. If we exist in time once we enter the afterlife, then what we could be seeing is some form of time anomaly associated with an event. Since we are from this reality, there could be a connection to an event that transcends the boundaries between the two realms. We can refer to the quantum entanglement effect where if two particles have a common origin and one changes then the other would undergo the same change no matter the distance. Since the body and spirit have a common origin then the two universes would be interconnected. It could be an effect like standing on the edge of a lake – you can see the fish swimming in the shallow water but once they swim too deep they're out of sight. We might be seeing the shallow end of time. If the entity involved in the residual type occurrence had no sense of time then the time factor involved would be inconsequential.

In another instance investigators went to a house located in New England, where there were claims of shadow figures seen in various parts of the house. One story told of a figure that could be seen looking in a window in the kitchen area. Another told of a figure sometimes seen in a doorway on the main floor. At certain times feelings of uneasiness would overcome those who were staying in the house. Pictures would fall from the wall in the basement. In the bedroom, a woman was seen standing by the bed and on one occasion a man wearing a hat was observed standing in the doorway. Cold spots could be felt in some areas and a man reported feeling a cold wind brush by him.

During the investigation, a team that was working upstairs could hear a banging noise coming from a bathroom, like someone knocking on the wall. After subsequent investigations they couldn't locate a source of the noise. While in the basement, investigators heard another knocking sound along with the sound of footsteps.

The team sat down and started an EVP session. They asked the usual questions such as asking for a sign of their presence. After a few minutes there seemed to be a response to their questions. They asked if the spirit could knock three times, and they'd get three knocks in response. They asked if the spirit could move an object or do something else to show that it was there. A closet door closed on its own, and after a few minutes the ceiling fan started to turn. The team then asked if it could show itself. A team member then saw a faint flickering at the end of a couch. The member said it looked as if something was trying to manifest but couldn't find the right frequency. The team had a thermal imaging camera and trained it on the couch area where the flickering had been seen. The camera showed faint heat impressions as though someone had been sitting there, although no one had.

Upon review of the evidence, they had caught an EVP from that time that said "I will do this" after asking the entity to show itself. That was the same time that the flickering effect could be seen. This has every indication of an intelligent haunt. The ghost or spirit responded in a positive way to prompts by the investigators. There is every indicator that this spirit was there and active. The reason this entity was there will probably never be known, but something brought it to that location. It could be attached to someone or something there. Intelligent attachments are not that uncommon where hauntings are reported. What causes them would be anyone's guess, but the fact that these spirits are there and interacting with humans in the present needs to be studied and possible answers found.

When a spirit begins an interactive haunting, I don't believe that it takes up permanent residence at the location. I would think that it visits whatever it is attached to as opposed to lingering there for all time. This would explain periods of activity and inactivity that occur. At the ruins of a castle estate in Ireland, investigators were looking into reports of paranormal activity. The estate was built in the 1690s and was in total ruin. All that was left of the original building were walls and part of a roof. Whatever was causing the activity must be very old. According to reports, voices could be heard and footsteps echoed in the ruins. At one time a couple of centuries ago, ancient Egyptian relics were kept there. It is thought that this might be the cause of the disturbances.

As the investigators worked into the night occasional voices and humming could be heard coming from no apparent area. As they entered a lower chamber they could hear the sound of boots on a wooden floor. The problem is that there wasn't any wood left – it had rotted away a long time ago. The sound of footsteps was recorded on digital voice recorders so there is credibility to the event. There were only two or three steps but there was no question as to what they were. The other noises and whispers were vague and distant but these were clear.

The sound of footsteps is a common occurrence during

hauntings, and I've often wondered how they could occur. As I have mentioned, ghosts seem to have the ability to alter the density of their form, even alternating from matter to energy because the difference between the two is the density of particles. If they concentrate density to a semi-solid form, then they could make footstep sounds. If they condense to a form they are familiar with — such as someone wearing boots — you have the sound of boots as footsteps. The problem is that they sounded like they were walking on wood or a wooden floor where there was no wood. The footsteps had to have been made in the past or the entity was in its past presently. Confused? Try to imagine this: if the entity is in *its* past it can also be in *our* present. The entity exists in time so it can be in any time it wants. Like the woman in the train station waiting for the train in the 1940s but relating the story to someone here, now. They can be there because they exist in time, but we don't so we can only experience our present. It is hard or even impossible for us to imagine moving through time as easily as we move through space and vice versa. It is difficult to imagine their ability to go from now to then. The fact of the matter is the footsteps were heard and recorded. They were on a wooden floor that wasn't there. This seems like a viable explanation.

There is an inn located near Gettysburg, Pennsylvania, there are stories of witnesses seeing the ghostly scene of doctors performing surgery on a patient lying on a table in a basement kitchen area that was used as an amputation hospital during the

Civil War. The doctors are said to look over to those watching and the scene just fades away. Apparently the severed limbs were thrown out a back door and the pile was high enough to block the window. Upstairs in one of the bedrooms, people have felt their legs being tugged while they sleep. In another room, footsteps could be heard and a rocking chair by a window is seen to move on its own. A shadow figure could also be seen in the hallway.

Investigators set up their equipment and proceeded to gather evidence of any paranormal event. While on the third floor, they could hear creaking noises and what sounded like footsteps made by someone wearing boots, possibly a Confederate soldier. One investigator was sitting on a couch and it seemed as though someone sat down beside him. He could feel the cushion depress. As they were sitting there the air became noticeably colder. A team was in the basement hoping to observe the surgical anomaly, but was unsuccessful. They did get an EVP after asking if whoever was there missed their family. They got a faint reply of "Yes" that only the recorder picked up.

As the investigation progressed the team actually spent the night there. In the room rumored to have the moving rocking chair, one investigator set up a camcorder pointed in that direction. Later in the night a noise could be heard. It was thought that maybe a candle on the mantle had moved but upon reviewing the tape a picture on a table by the rocking chair could be seen to move as if someone was sitting in the chair and turned the picture as if to look at it. It was a very nice piece of evidence.

So what we have at this location is what seems to be an intelligent haunt because of the response to EVP questions and there definitely is a presence there because of the picture turning. There also is what some would call a residual haunt due to the occasional occurrence of the surgical scene. As before, I believe that this would be more along the lines of a time anomaly as opposed to memories implanted somehow. What we have been doing is trying to understand and offer possible explanations for various types of hauntings. Hauntings occur. That is a fact. Once again, I call on scientists to look at this with the intent of discovering the nature of our next existence.

CHAPTER 12

The Case for Scientific Inquiry

I really believe that some kind of valid scientific study needs to be done to try and understand exactly what is going on concerning our afterlife. The existence of parallel universes is a probability according to current research and hypotheses in the field of physics, and it seems as though spirits observed are associated with time and time periods. We need to know why some spirits are here and others are not.

What we call death is not the end. A number of noted scientists and atheists have stated publicly that there is no afterlife. Physicist Stephen Hawking, for example, has stated that before the Big Bang there was nothing. Since there was nothing – no space, no time – then there was no God to create anything, therefore there would be no heaven or hell to which a soul or spirit would go. Hawking says that when we die there is nothingness, life ceases absolutely, the end.

Other physicists including Michio Kaku have offered a rebuttal that with the advance of string theory, the universe had no beginning and will have no end. Time has existed forever.

They say that the universe is actually a multiverse consisting of infinite parallel universes and that Big Bangs occur all the time over infinity. The existence of parallel universes is becoming an accepted fact in the realm of physics as more and more is learned about strings.

I am not a physicist, but I find convincing the hypothesis that the Big Bang resulted from the explosion of a white hole, which is the opposite end of a black hole. I believe that the wormhole created by the black hole in a parallel universe is still there and that particles and space continue to enter this universe through it. The pouring in of space would account for the expansion of the universe. When some cosmologists consider the Big Bang they fail to account for the source of material. They can account mathematically for the singularity — the core matter from which everything came — but they don't take into account the probability of such an occurrence. They might be able to prove mathematically that all of the world's oceans could fit into a tea cup, but the probability of that happening is highly suspect. They don't take into account the source of material.

Not only did energy and matter appear after the Big Bang, space also was created. Scientists really don't know what space is. If you look up space in an encyclopedia it says that space is one of the few fundamental quantities in physics, meaning that it cannot be defined by other quantities because nothing more fundamental is known about it. In other words, they don't know what it is. It can be bent, affected by gravity, and is expanding at an incredible rate. There is no other hypothesis about the origin of space except the Big Bang. There again, there is no accounting for the source of material. Space is not particulate, so it is not observable in a known sense. If space, which is something in itself, is pouring in through a massive wormhole connecting two parallel universes, then the source and expansion could be explained.

Einstein's theory of relativity states that space and time are variations of the same thing. When space pours in, time flows in also. We have what Einstein called space-time. From this, we can infer that space and time exist in some shape and form in most if not all parallel universes. There are parallel universes, and these universes can interact with each other. There are an infinite number of parallel universes, and when we become energy beings then we can exist in one or more of them because physical laws in those environments would allow this.

When Hawking made his comments about the impossibility of an afterlife, he stepped briefly into the world of the paranormal. This is an area dismissed as pseudo-science. Hawking used the principles of physics to express his opinions about the possibilities of life after death. I have used the principles of physics to express mine. Scientists don't study the paranormal because there is no core of information available such as in physics or astronomy, etc., on which to base their studies.

They don't have the building blocks laid down by centuries of trial and error and meticulous observation. They study their various sciences in colleges and universities to obtain the knowledge collected over time so they have foundations to expand upon. They don't want to apply their expertise to an area where there is little provable data. They base their entire science on the observable realities of the known universe. They lack a definite starting point and so they shy away from paranormal phenomena.

There are amateur investigators using whatever equipment they can to gather information needed to prove the existence of ghosts. These investigators are becoming very resourceful in their endeavors. The legitimate ghost hunters are accumulating evidence such as pictures, videos, voice recordings, and personal experiences that they present to the world as authentic. We have to differentiate between the evidence presented by serious individuals and the fake and misrepresentative so-called ghost videos and pictures on the internet.

I am not aware of any significant scientific research into ghostly or paranormal activity. I recall researchers setting up cameras and recorders at Eastern State Penitentiary in an attempt to capture paranormal activity. I don't believe they caught anything so the conclusion was that this type of activity can be explained away. As a deer hunter, I've spent many days sitting in the woods without seeing any deer while I was hunting, but I don't then conclude that deer simply do not exist.

The outdated thinking that the study or investigation of ghosts is taboo and should be ridiculed needs to be changed. We have mapped the human genome and traced our origins back as far as we can, so far, and think we have a complete picture of our existence at hand. We will never have the complete picture until we understand or accept what happens after this life is completed.

CONCLUSION

Going back to the beginnings of human existence, we have seen evidence of ritual burial from our earliest days. Belief in an afterlife was part of the fabric of virtually every culture throughout history. Our beliefs in ghosts go back just as far. Myths and suppositions about ghosts have fascinated mankind for thousands of years. The idea that ghosts are goblins in the night and should be feared needs to be replaced by knowledge and understanding. If we know what goes bump in the night then we shouldn't be afraid of it. After all, ghosts are just people without a body, but the primal fear has been instilled in us regardless.

As we have looked at the history of how different cultures rationalized and prepared for the afterlife we have seen that the same themes are presented in different ways. The premise that our spirit carries on is always the basis, but religion and cultural differences influence concepts of what awaits us after we cross over.

I have attempted to lay the foundation for legitimate scientific inquiry. If we can apply the basic scientific principles using known principles of physics in an attempt to understand the science of the afterlife, we can begin to complete the human equation. The fact that we can use science and known principles to explain and understand what happens in a second or after life is encouraging. The application of these principles adds weight to the possibility of continual existence. If there were no explanation at all for these events or if the science and laws of science could not possibly account for this phenomenon, then I would say that they are right and this does not exist. But with proper investigation and examination we may actually find an answer to our questions about the nature of our existence, or at least the groundwork for an explanation.

It is possible that there could be different incarnations of afterlife. Once we become energy or particulate after our organic existence, we could evolve to a spiritual or pure intelligence form and more beyond that. The prospects are limitless. There are indications that when we pass, we exist in time, probably in parallel universe with its own set of physical laws. I have said that when I go, I want to visit the '60s and listen to some old rock 'n roll – maybe jam with Jimi Hendrix or something.

We have examined the possibility of the existence of god. If we exist in time and the past will always exist per Einstein's theories, then it would be the perfect way for us to exist forever. To me that is evidence that there is a divine presence — maybe not like the one professed by the clergy, because organized religion has been manipulated by power and politics for thousands of years — but the essence of intelligent design. The facts are there if you look. Ghosts exist. I have pictures and experiences to prove it. There is an afterlife. Something happens and I have tried to offer some kind of explanation using my observations and the observations of others as to what is happening. Maybe I am wrong, but I leave it up to legitimate scientists to prove that and in doing so find true answers to complete the picture of human existence. We actually have a name for this parallel universe we enter into upon passing. It's called Heaven.

EPILOGUE
Ghost Hunting 101

I encourage those of you who have the need to know to go on your own ghost adventures with some basic equipment and find out for yourselves that this is real and observable. You will need an EMF detector, which you can get online fairly cheap, a camera (I use disposable 35 millimeters because they seem to capture more than digitals) and a voice recorder, which can be obtained at many discount stores. Check out locations that offer organized ghost hunts. I have found that areas that are said to be haunted usually are. Have some fun, get some thrills and see for yourself that this is real.

Ghost Hunting Equipment

I have a Bell & Howell night vision digital camera and a Medtronic infrared digital camera for which I paid $130 each. Everything else that I have would be considered "consumer grade" with each item costing $99 or less, since I am a person with limited means. If you have the need or desire to purchase such equipment, there are several outlets to consider. The one I have used the most often is eBay. All you need is a Paypal account to make a purchase.

These are some of the items in my bag of tricks.

- Bell & Howell Night Vision Digital Camera

    This is my favorite piece of equipment. I have caught a lot of ghosts with this from orbs to shadow figures to full-bodied apparitions. Easy to use, fairly inexpensive, and I consider a must-have. Takes pictures in complete darkness.

- Medtronic Infrared Digital Camera

    This was my first specialty camera. This camera shoots in the infrared spectrum. Regular digital cameras have filters to filter out unwanted spectrums of light to produce pictures we are accustomed to. These are also called full-spectrum cameras. The pictures produced come out with a reddish or bluish hue and can pick up more than a traditional digital camera.

- Disposable 35 millimeter camera

    These can be bought at any Walgreens for less than $10. My personal experience with this type of camera is that they seem to capture ghostly images better than digital cameras. The drawback to this type of camera is that the film has to be developed instead of uploading the images to a computer via a media card or USB cable, but the modest expense of film processing is worth the added ability to capture evidence, which is the name of the game. I have literally dozens of ghost and paranormal pictures and quite a number of them were taken with a disposable camera. Give it a try.

- Digital Outdoor Trial Camera

This is the type of camera hunters use to track a deer. It is attached to a tree and is motion activated. It uses a media card like a traditional digital camera. I picked this up at a deer and turkey expo for $49.99 and thought it might be a nice addition to the arsenal. I use it in secluded rooms while I am investigating elsewhere. I have actually caught a couple of strange things as they pass in front of the camera. This makes an excellent camera trap for light or dark situations as it has an infrared feature.

- Digital Camera Binoculars

What an idea — putting a digital camera in a pair of binoculars. These use a media card like a standard digital camera and can take distance pictures. I have had an occasion to use them. I wanted to investigate a Civil War-era mansion about 50 miles away from where I live, but the owners don't let allow anyone on the property. I got as close as I could and used my camera binoculars to take some pictures. I caught what looks like a shadow figure in a front window. This is not something I use all the time, but it was there when I needed it.

- Laser Digital Thermometer

    This is a hand-held thermometer with a laser-pointer feature. You point the laser and it reads the ambient temperature. This is excellent for locating cold spots

- Motion Sensor

    A battery-operated motion sensor lights up when something passes in front of the sensor. This can be used in conjunction with a camcorder to document movement of a shadow figure or entity.

- "Flip" Night Vision Camcorder

    Flip cameras are inexpensive compact camcorders used for surveillance work. I bought mine at Kmart for less than $40. Place one on a tripod in a hallway with the motion sensor and see what you catch. This uses a standard video media card and can be played back on a computer.

- Laser Grid

    This is a really cool toy. A device that looks like a pencil fits on a bipod. The bipod actually separates the on and off switch. A grid shines points of light dispersed over a large area, and if anything moves through the area points of light are blocked out. This is very good for finding shadow figures. You can use it with a flip camera and motion detector to set a more elaborate trap.

- Voice Recorders

I have a couple of types of voice recorders. Compact digital voice recorders can be found at Walmart or electronics stores for about $40. I also have compact voice recorders that use mini-cassettes. I actually bought a couple of them at a flea market. Voice recorders are used to record EVPs, or electronic voice phenomena. EVPs are voices picked up by the electronic equipment but not heard by the human ear. Investigators ask questions in a haunted location and the response is recorded but not heard. A recorder can be placed where ghostly footsteps or other strange sounds are reported. When I use a voice recorder, I prefer the old-style mini-cassette recorders over the digital because I think they pick up more detailed sounds. This is another strike against the digital.

- Flashlights

You may think a flashlight is a flashlight is a flashlight, but this is not true. I have a flashlight with a laser pointer in the center of the beam that I can turn on by itself. I use this in a hallway or room and watch the beam to see if anything crosses in front of it. It's like the laser grid but with a single point of light. The flashlights I prefer are the ones that clip onto the bill of a hat, which makes them hands-free. I use either green or red light because these colors don't interfere with night vision. Hat lights can be found at any sporting goods store.

- Night Vision Goggles

Yes, I have night vision goggles. Believe it or not, I found them at a toy store for about $50. They look just like the ones the military uses, but they are made of plastic. They work.

- Static Electricity Detector

This is a small box with a short antenna that senses static electricity. Theory has it that ghosts are made up of energy and this device senses that. It has a light that comes on when activated and the light goes off when energy is present. More toys for the trap.

- Spirit Box

This AM radio also is known as a ghost box, and scans all of the channels in a matter of seconds. What this does is create white noise that many believe ghosts or spirits can use to form or mold words. You would use this when using voice recorders during EVP sessions.

- EMF Detector

I actually have four EMF detectors. EMF stands for electromagnetic field. These detectors originally were used by electricians, but now are a staple in the ghost hunter's toolbox. The first one I have is a **Gauss Master Detector**. It has a meter that gauges the strength of the field detected. Since ghosts are thought to be made of energy, a reading well above the base reading of the area could well mean the presence of a spirit. This meter also gives an audible sound when high EMFs are detected. We use this audio signal as an opportunity to take a picture. I have caught quite a few ghosts on film this way.

The second detector I have is along the same lines as the first, the difference being that it's digital and doesn't make a sound. This one can be turned on and used with your motion detector and flip camera so in the event the motion detector activates, you can get an EMF reading to confirm a presence. This is the EMF detectors used on *Ghost Hunters*.

The third one is the **K2 meter**. This type of meter flashes lights when an electromagnetic field is detected. The more lights, the stronger the field. Some investigators use the K2 to try to communicate with a ghost. Investigators ask questions that require a yes or no response, and the ghost is asked to illuminate so many lights for yes and so many for no. You also can add this to your trap. The price range for these devices is usually between $40 and $60.

The fourth EMF detector I have is the **MEL meter**. This is so far the most advanced meter on the market. It detects

EMF in A/C (alternating or household current) and D/C (direct current). It also interacts with the Earth's natural geomagnetic field. This unit also registers temperature and records readings. This is the one favored by the *Ghost Adventures* investigators. I bought my tester for $99 but think the price has risen since then.

- EM Pump

  EM or electromagnetic pump creates an electromagnetic field. It is actually an EM generator. A small box with a simple on-off switch generates a field that many believe ghosts use to try to manifest or communicate. The EM pump can be used along with the ghost box during an EVP session in an effort to communicate with a spirit. Do not use an EM pump along with EMF detectors in your trap because the field generated by the pump will give false readings. EM pumps are usually in the $30 price range.

Places to shop online

- http://www.ebay.com — Online auction site where new and used equipment often can be purchased at a bargain.
- http://www.nbcuniversalstore.com — Parent company of the SyFy Channel has an online store where officially licensed Ghost Hunters TV show products are sold,

including ghost hunting kits and tools.

- http://theghosthunterstore.com/ — Online store run by New Jersey ghost hunters Michelle and Dave Juliano offering equipment they use in the field.
- http://www.ghostaugustine.com — Outlet offering ghost tours in St. Augustine, Fla., also has an online store with bargains on ghost hunting gear.
- http://www.amazon.com — Online marketplace for pretty much everything.
- http://www.ghoststop.com — Small business started by a paranormal investigator based out of Florida that sells assorted ghost hunting gear used on shows like Ghost Adventures, Ghost Hunters International and Fact or Faked.

These outlets specialize in specific ghost hunting equipment such as EMF detectors and EM pumps, etc. Generic items such as flashlights, digital voice recorders and digital cameras, disposable cameras, and non-ghost specific items can be bought fairly reasonably at Walmart, or ordered through some of these websites. You can spend as much as you want or as little as you need depending on what kind of evidence you want to collect and how technical you want to get. I must warn you that ghost hunting is very addictive and the related toys become must-haves.

Haunted Locations

One thing I have learned is that if a place is rumored to be haunted it usually is. If you want to do some ghost hunting yourself, there are many places that offer the opportunity. I would offer some advice to those who want to take the plunge and try to gather ghostly evidence. Go to places that offer ghost hunting opportunities such as old prisons or historical locations. A lot of these places do so as fundraisers for restoration programs. I would recommend not using Ouija boards or conducting séances at any time. The reason for this is that they can invite demonic entities. Avoid areas that could possibly be inhabited by demonic forces. Demonic energies never walked the earth as humans. They are a form that has evolved elsewhere. If you are haunted by a demonic entity, they can follow you if you move and are difficult to get rid of. Established ghost hunting locations usually are safe. You can gather great evidence and take excellent pictures while investigating at these places.

Gettysburg

One place I believe every ghost hunter has to visit is Gettysburg. This could be one of the most haunted areas in the United States. During the battle, more than 51,000 Civil War soldiers lost their lives. There are numerous locations from private residences to open fields that have ghostly encounters associated with them. One location with paranormal activity is the George Weikert house. Some of the stories associated with this location are of a door on the second floor that won't stay closed and footsteps said to be heard walking back and forth in the attic as if someone was pacing in anticipation of an upcoming event.

Another location that has gained fame is Sachs Bridge. This covered wooden bridge was used as a makeshift hospital to treat the wounded during the battle. There have been EVPs captured there as well as photos of what seem to be soldiers walking across the bridge. On rare occasions the sounds of gunfire and cannon barrages can be heard along with flashes of light in the night.

There is an area where a battle took place on the second day of the historic event called the Devil's Den. This fight took place in an already haunted place. In times before, there was a Native American encounter there called the Battle of the Crows when many lives were lost. After this there were legends of paranormal activity there. During the Civil War, this was a battle that saw many men killed, and turned into one of the defeats for the Confederate forces. The carnage there was horrific. Witnesses described the dead and dying lying everywhere. Many weren't buried for days or weeks and some were just thrown in crevices between the rocks. This onslaught would have left much spiritual energy. It is said that the sounds of battle can still be heard, and the pungent odor of gunpowder lingers in the air.

During the battle known as Pickett's Charge, more than 12,500 Confederate soldiers attacked Union forces crossing an open field trying to fight uphill toward Union lines. This turned into a scene of massive carnage as the Yankees fired volleys of artillery toward the advancing Confederate forces. Because of a lack of communication and tactical blunders, the Confederate Army suffered a grueling 50 percent casualty rate, while the Union army lost 1,500 men. There were many lives that were taken prematurely and suddenly — the kind of event that would leave many souls searching for closure to their brief lives. Many perished with unfinished business, which I'm sure would account for the many apparitions that inhabit the battlefield. At times columns of soldiers can be seen maneuvering as if still in battle mode. These time anomalous visions have been observed several times. Having no perceived sense of time, the spirits involved will carry out their enactments probably forever. With their ability to exist in time the possibility exists that maybe in some reality the outcome of the battle and even the war could be different. That could be what they are actually doing. Existing in time all possibilities exist.

- For a listing of ghost tourism opportunities at Gettysburg, go to http://www.gettysburg.travel/visitor/gettysburg_ghosts.asp

Ohio State Reformatory

One of my favorite places to ghost hunt is the Ohio State

Reformatory in Mansfield, Ohio. They conduct ghost walks, which are tours of paranormal hot spots lasting several hours, and they have ghost hunts lasting all night, in which you have run of the place and can go almost anywhere in the prison. The Reformatory was built in 1886, but it wasn't until 1896 that the first prisoners were incarcerated there. The first prisoners there were put to work finishing the construction and the East Cell Block was not completed until 1908. If you haven't seen pictures of it or haven't been there, the place is evocative of Dracula's castle. It is a very cool-looking old building and the restoration projects continue with the help of money generated by the ghost events. The site has been the set for several movies over the years. *Harry and Walter Go To New York*, *Tango and Cash*, *Air Force One*, and *The Shawshank Redemption* were filmed there.

Kane Hodder, the actor who played Jason in the Friday the 13th movies, was filming the movie *Fallen Angels* at the Reformatory when he and a stunt man decided that after filming for the day they would go and do some ghost hunting. After a while, they saw a black shadow figure in one of the cell blocks. After chasing and losing sight of the figure, they returned to California and formed their own paranormal group – all of this because of what they saw at the Ohio State Reformatory.

There have been many deaths at the Reformatory. In 1926, a corrections officer was shot and killed in an attempt by a former prisoner to break out another prisoner. In 1932, while in what's called "the hole" or solitary confinement, another officer was beaten to death with an iron pole. There have been numerous inmate murders and suicides. At one time a warden's wife was shot and killed in an accident. Nine years later, the warden whose wife had died suffered a heart attack in his office and passed away there. Deaths and hardships continued until its closing in 1990. After the closing, visitors have reported hearing cell doors close and many strange noises. There are footsteps in the cell block when no one is there and black shadows can be seen entering and leaving cells in different parts of the prison. At times, people have reported being touched or their hair being pulled.

I was there once for a Halloween event during which they have a Halloween haunted prison set up in a haunted prison. They had a closed circuit television camera set up showing a stairway where a black figure was reported to have been seen. One of the guides said that he had been punched while provoking spirits in the shower area. I guess they didn't like that much. As far as provoking goes, I really don't use it while ghost hunting because I believe the spirits will respond better if you treat them with some respect. I know I would.

In the warden's office area, there is a shadow figure seen that is believed to be the warden's wife who was accidentally shot. The figure is said to move gracefully through the area. Another figure reported could be the guard killed in the hole area.

He could still be there watching over the prisoners there. At times you can hear the jingle of keys as if he was making his rounds. As you walk through the cell blocks whispering and voices sometimes emanate from empty cells. Attachment to certain things could somehow overcome the pleasure of being where you're supposed to be when you pass. To some sentenced there, it could be the closest thing to an actual home that they might have experienced and are afraid of losing that even as bad as it was. Perhaps like the characters in *The Shawshank Redemption*, they environs of the prison have become so familiar they no longer know how to exist outside its crumbling walls.

- Ghost tours and hunts at the Ohio State Reformatory are organized by the Mansfield Reformatory Preservation Society. Information is available on the group's website at http://www.mrps.org/

Eastern State Penitentiary

When we think of haunted prisons we can't overlook Eastern State Penitentiary. It was built in Philadelphia in 1829. It was constructed by the Quakers as an experiment in punishment. Their idea was to put prisoners in isolation so they could achieve penitence with God. The word penitentiary is derived from this. Before Eastern State, the word penitentiary didn't exist.

The prisoners were put in cells alone and were only allowed outside into a very small courtyard briefly every week.

They had no human contact except for limited contact with guards and ministers who would pray with them. The effect of isolation was the opposite of what the prison's creators had expected and desired. Instead of spiritual enlightenment and penance, the result was insanity and deprivation. The basis for this method of incarceration was the concept used by monks in Europe. The Quakers thought that if the monks could achieve penitence and peace through isolation and silence then the prisoners could also. They failed to recognize the discipline involved.

One of the most famous residents of Eastern State was Al Capone. Capone was arrested in Philadelphia along with bodyguard Frankie Rio for carrying concealed weapons. They were sentenced to a year in prison at Eastern State. During his time there Capone was reported to be haunted by the ghost of James Clark, who was one of the St. Valentine's Day Massacre victims. It is said that Capone could be heard screaming in his cell and begging Clark to go away and leave him alone. Even after his release, Capone thought that Clark's ghost was haunting him.

Cell Block 12 reportedly is one of the prison's paranormal hot spots. At this location, witnesses over the years have reported hearing distant laughter echoing through the cell block. Could it be the laughter of some insane tenant tormented throughout time? There also have been reports of shadow figures vanishing into different cells. On death row, there have been strange encounters with dark apparitions and feelings of dread.

There was a locksmith working on one of the locks leading from a courtyard to a cell when he had a strange encounter. He felt as though he was being watched but he looked around and no one was there. He continued working when the feeling again became overbearing. He quickly looked around and saw a black figure standing by him. As soon as he saw it, the figure darted away at amazing speed.

Another ghostly figure sometimes seen is that of a specter seen in a guard tower above the prison wall. It is thought this might be the ghost of a former guard who still is watching over the place. I guess if there are still prisoner spirits about, then someone has to oversee them.

An investigative team was in a cell block trying to catch a ghost at play when one of their cameras caught what looked like a cloaked figure run down the catwalk. Upon reviewing their evidence, the figure could be plainly seen and it looks as though it materializes in front of the camera before it runs away. It was dark and came out of the darkness. I'm sure it was one of many ghosts lingering there.

- For information about visiting Eastern State Penitentiary, go to http://www.easternstate.org

The Myrtles Plantation

The Myrtles Plantation in St. Francisville, Louisiana, was built by David Bradford in the late 1700s. He named the plantation Laurel Grove and lived there with his wife and five children. One of his daughters, Sara Mathilda, married Clark Woodruff, who managed the plantation for his mother-in-law following Bradford's death. Under his management the plantation was expanded to produce indigo and cotton. Woodruff and Sarah had three children: Cornelia Gale, James and Mary Octavia. Tragically, Sarah Mathilda died after contracting yellow fever on July 21, 1823.

After his wife's death, Clark continued to maintain the plantation with the help of his mother-in-law, Elizabeth. The tragedy continued, though, with his son James dying from yellow fever on July 15, 1824, and then in September of that year he lost his daughter Cornelia Gale to the same dreaded disease. Elizabeth continued to live there with Clark and Mary Octavia until her death in 1830.

In 1834, Laurel Grove was sold to Ruffin Grey Sterling. He and his wife Mary Cobb did extensive remodeling and construction projects at the plantation. He extended the walls of the original house to create four larger parlors, a formal dining area and one that was used as a game room. The finished structure was almost double in size from the original, and he changed the name of Laurel Grove to the Myrtles. It was only four years after the remodel was complete that Ruffin Sterling died of consumption. He left the Myrtles plantation to his wife, who managed it for many years.

Through the years, tragedy followed the children of Ruffin and Mary Cobb. The oldest son, Lewis, died in 1854, the same year his father passed away. Daughter Sarah's husband was murdered on the front porch after the Civil War. Sarah continued to live there until her death in 1878. The property was purchased by Stephen Sterling and being deeply in debt he sold it in 1886 to Oran D. Brooks, who held the property until 1889 when Harrison Milton Williams bought the estate. The Williams family kept possession until the 1950s when the property and assets were divided among heirs of the Williams family and the house was purchased by Marjorie Munson.

It is said that after Munson bought the house stories of ghostly encounters began. One of the encounters being reported is that of a woman wearing a green bonnet or headdress. She is described as being an older black woman and has been seen standing by a bed holding a candlestick. This entity has actually been photographed. This apparition could be that of one of the many slaves who endured existence there before the Civil War.

Another specter being reported is that of a Confederate soldier being seen by a pond adjoining the living quarters. Although no battles were recorded there, a skirmish or encounter with Union forces could account for the soldier ghost seen there.

There are reports of ghostly children being seen and heard in various areas of the house. The young boy and girl supposedly occupying the residence could be those of the Woodruff children, who were victims of a yellow fever outbreak.

Apparently the grand piano located on the main floor plays by itself. While filming location scenes for a miniseries based on *The Long Hot Summer*, crew members re-arranged furniture in several rooms and then left. Upon returning they found that the furniture had been placed back in their original positions. Spooky stuff. The Myrtles is open for guests and they do ghost tours. It's on my list of places to go.

- For information about tours, go to the plantation's official website at http://www.myrtlesplantation.com

Ax Murder House

For those of you who are not faint of heart, there is the Ax Murder House in Villisca, Iowa. At this location, eight people were brutally murdered while they slept. The victims included two adults and six children. Josiah B. Moore, his wife Sarah, and their children Herman, Katherine, Boyd and Paul, along with their friends Lena Gertrude and Ina May Stillinger died at the hands of an unknown assailant on June 10, 1912. There has never been an arrest or conviction associated with this crime.

It is said that the murder or murderers entered the house

and waited in the attic for the family to return home from a church outing. There were cigarette butts on the floor of the attic as the murderer or murderers waited for the ill-fated innocents to fall asleep. Sometime during the night they emerged from the attic area and systematically hacked each one to death with an ax. The estimated times of death were shortly after midnight.

All of the curtains in the house were drawn except for two windows that didn't have curtains, but were covered with clothing belonging to the victims. It is in the report that all of the victim's faces were covered with bedclothes after they were murdered. Kerosene lamps were found at the base of the bed in several rooms. An ax assumed to be the murder weapon was found in the room where the Stillinger girls were sleeping. The ax is thought to belong to Josiah Moore. One of the gruesome finds was gouge marks in the ceiling caused by the swinging of the ax.

Whenever the attacker left, he or she locked the door behind them because all of the doors were locked when authorities arrived. A neighbor, Mary Peckham, was the first to notice the inactivity at the Moore house. She called Mr. Moore's brother, who came over and discovered the murder scene.

Even though there was never an arrest, there were several suspects in the murders. One of those was Frank F. Jones. Jones was a prominent citizen and state senator. Josiah Moore had been employed by Mr. Jones when he decided to start his own business. As it turned out, Mr. Moore took one of the best accounts with him into the new business venture, which infuriated Jones. Additionally, there was a rumor that Moore was having an affair with his daughter-in-law.

Another suspect was Reverend George Kelly, who many considered to be crazy. Another theory is that they were targeted by an unknown serial killer. Whoever did the deed left a brutal mystery that might never be solved.

One of the stories of hauntings occurring at the house claims that the spirit of the killer is there. Several psychics and mediums have said they feel the presence of the murderer there. Even though the killer didn't die there, the experience of the murders could have left such a profound impression on him that once he passed away his ghost would be compelled to visit the crime scene.

There are reports of black shadows and masses encountered in the house along with voices and EVPs said to record the ghostly voices of the murder victims. Witnesses report doors that open and close by themselves, cold spots, and feelings of dread. I don't think anything demonic has been reported to reside there, but there are reports of feelings of evil and an evil presence. Like I said, anyone who would like to attend a ghost hunt here should not be faint of heart or frighten easily, but should have some kind of understanding of what happens when we die. Knowledge would be the best weapon here.

- Both daytime tours and overnight stays are offered at the

Villisca Ax Murder House at http://www.villiscaiowa.com/

Other hot spots to check out

- Waverly Hills Sanitarium, Louisville, KY — a hospital where thousands of tuberculosis patients died over five decades. Information at http://therealwaverlyhills.com
- Iron Island Museum, Buffalo, NY — the site of a former church and funeral parlor in the Iron Island neighborhood of Buffalo. Numerous reports of paranormal activity have been made there. The museum offers public tours and overnights. Information at http://www.ironislandmuseum.com/
- Prospect Place Mansion, Trinway, Ohio — historic mansion that once was a stop on the Underground Railroad. Tours and ghost hunts are offered. Information at http://prospectplace-dresden.com/
- Whisper Estate, Mitchell, IN — old Victorian home with lots of reported paranormal activity. Tours and ghost hunts available through http://www.whispersestate.net/
- Morrison Lodge, Elizabethtown, KY — Masonic lodge and former Civil War field hospital.
- Bobby Mackey's Music World, Wilder, KY — haunted country music night club featured on *Ghost Adventures*. Information at http://www.bobbymackey.com/
- Stanley Hotel, Estes Park, CO — This inspiration for Stephen King's classic horror novel *The Shining* has a reputation as

one of the most haunted spots in America. The hotel remains open and reservations can be made at http://www.stanleyhotel.com/

- Fort Mifflin, Philadelphia, PA — A fort dating back to the American Revolution with a lengthy history of paranormal activity. http://fortmifflin.us/paranormal/
- Shanghai Tunnels, Portland, OR — A network of underground tunnels where men were kidnapped into serving on ships and women were sold into prostitution for decades beneath Portland's streets. http://www.shanghaitunnels.info/
- West Virginia State Penitentiary, Moundsville, WV — Former prison with lots of paranormal hot spots. http://www.wvpentours.com/
- Trans Allegheny Lunatic Asylum, Weston WV — Former lunatic asylum dating back to the mid-1800s. http://trans-alleghenylunaticasylum.com/
- St. Augustine Lighthouse, St. Augustine, FL — A 19th century lighthouse that stands entrance to the first European settlement in the United States dating back to the 1500s. http://www.staugustinelighthouse.com/
- Preston Castle, Ione, CA — Former juvenile reform school offering overnight ghost tours. http://www.prestoncastle.com/
- Birdcage Theater, Tombstone, AZ — Old theater and saloon in notorious the notorious Old West town of Tombstone. http://tombstonebirdcage.com/
- Lizzie Borden House, Falls River, MA — Home of legendary

accused ax murderess Lizzie Borden, now a bed and breakfast. http://lizzie-borden.com/

BIBLIOGRAPHY

Books

Belanger, Jeff. Encyclopedia of Haunted Places. New York: Castle Books, 2005.

Deutsch, David. The Fabric of Reality. New York: Penguin Books, 1997.

Hawking, Stephen. A Brief History of Time. New York: Bantam Books, 1988.

Kaku, Michio. Parallel Worlds. New York: Anchor Books, 2005.

Musser, George. The Complete Idiot's Guide to String Theory. New York: Penguin Group, 2008.

Ouellette, Jennifer. The Physics of the Buffyverse. New York: Penguin Group, 2006.

Post, John F. Metaphysics: A Contemporary Introduction. New York: Paragon House Publishing, 1991.

Taylor, Troy. The Haunting of America. New York: Barnes & Noble Publishing, 2001.

Thay, Edric. Ghost Stories of Ohio. Edmonton: Ghost House Books, 2001.

Webb, Stephen. Out of This World. New York: Copernicus Books, 2004.

Online sources

"A 27,000-year-old burial site is studied." Phys.Org. http://phys.org/news82923130.html

Adams, Cindy. "An Ancient View of the Afterlife." http://www.gather.com/viewArticle.action?articleId=281474977417367

"Afterlife Beliefs and Phenomena" James Lewis' Afterlife Studies. http://www.near-death.com/religion.html

Ancient Egypt. http://www.allabouthistory.org/ancient-egypt.htm

Archeociel.com. Site of archaeostronomer Chantal Jègues-Wolkiewiez.

Bagwell, Kristina. "Burial Rituals and the Afterlife of Ancient Greece." https://staff.rockwood.k12.mo.us/materkristen/.../Burial_Rituals.pdf

"The Belief in Ghosts in Greece and Rome." Scary Stories.ca. http://www.scarystories.ca/GhostStory/The-Belief-In-Ghosts-In-Greece-A.html

"Brain Areas Critical to Human Time Sense Identified." Daily University Science News. http://www.unisci.com/stories/20011/0227013.htm

Dewey, Stephen. "Continuing popular belief in the supernatural in the 19th century." Myths & Mysteries, Predictions, Psychic Power. Sept. 1, 2004. http://www.skepticreport.com/sr/?p=164

"Double-slit experiment." Wikipedia.org. http://en.wikipedia.org/wiki/Double-slit_experiment

"The Einstein-Podolsky-Rosen Argument in Quantum Theory." Stanford Encyclopedia of Philosophy. http://plato.stanford.edu/entries/qt-epr/

Fram, Alan and Trevor Thompson. "That's the Spirit: Belief in

Ghosts High." The Associated Press, Oct. 26, 2007.

Gill, N.S. "Roman Burial Practices." About.com. http://ancienthistory.about.com/od/deathafterlife/a/RomanBurial.htm

"Ghosts." Encyclopedia of Death and Dying. http://www.deathreference.com/En-Gh/Ghosts.html

"Ghosts in Chinese Culture." Wikipedia.org. http://en.wikipedia.org/wiki/Chinese_ghosts

"Ghosts in Islam." http://www.angelsghosts.com/ghosts_islam

Heller, Amy. "Archeology of Funeral Rituals as revealed by Tibetan tombs of the 8th to 9th century." http://www.transoxiana.org/Eran/Articles/heller.html

"Homo Heidelbergensis." Wikipedia on Answers.com. http://www.answers.com/topic/homo-heidelbergensis-1

Kingsley, Danny. "World's oldest burial redated to 40,000 years." ABC Science Online.

"Leif Ericson and Vikings." http://www.vikingship.org http://www.abc.net.au/science/articles/2003/02/20/788032.htm

Mayell, Hillary. "Oldest Human Fossils Identified." National Geographic News. http://news.nationalgeographic.com/news/2005/02/0216_050216_omo.html

"Oldest Discovered Burial Site." Encyclopedia of the Unusual and Unexplained. http://www.unexplainedstuff.com/Afterlife-Mysteries/Oldest-Discovered-Burial-Site.html

Pihlajamaa-Glimmerven, L.E. "Biological Clocks."

http://glimmerveen.nl/LE/biological_clock.html

"Prehistoric Cave Paintings." Kathy Ceceri's Crafts for Learning. http://www.craftsforlearning.com/prehistoric.htm

"Religious Beliefs in Heaven and Hell." http://www.religioustolerance.org

Sandri, Danilo and Alessia Genito. "Ghosts…these mysterious presences." Analysis of Oscar Wilde's *The Canterville Ghost*. http://www.itiscannizzaro.net/Ianni/booksweb/canterville/beliefs.htm

Streich, Michael. "Ancient Tombs and Burial Practices." Suite 101. http://suite101.com/article/ancient-tombs-and-burial-practices-a101082

Walker, Cliff. "The Paranormal and the Supernatural, as Each Relates to Science." http://www.positiveatheism.org/mail/eml8958.htm

"What the Bible says about ghosts and psychics." http://www/angelfire.com/mi/dinosaurs/ghosts.html

"What Does the Roman Catholic Church Say About Ghosts?" Ehow.com http://www.ehow.com/about_4577988_roman-catholic-church-say-ghosts.html

"World's Oldest Jewelry Found." Science 2.0. http://www.science20.com/news/worlds_oldest_jewelry_found

www.ingramcontent.com/pod-product-compliance
Lightning Source LLC
Chambersburg PA
CBHW020903090426
42736CB00008B/483